Calculus II Exam File

D. R. Arterburn, New Mexico Institute of Mining and Technology, Editor; Bill Bompart, Augusta College; Peter Braunfeld, University of Illinois at Urbana-Champaign; William E. Demmon, University of Wisconsin Center - Manitowoc; Billy Finch, University of Florida; Joe Flowers, Northeast Missouri State University; Michael E. Frantz, University of La Verne; Biswa N. Ghosh, Hudson County College; LeRoy P. Hammerstrom, Eastern Nazarene College; John H. Jenkins, Embry-Riddle Aeronautical University; Ann F. Landry, Dutchess Community College; David H. Lankford, Bethel College; Eric M. Lederer, University of Colorado at Denver and Red Rocks Community College; John Martin, Santa Rosa Junior College; Varoujan Mazmanian, Stevens Institute of Technology; Thomas A. Metzger, University of Pittsburgh; Alejandro Perez, Laredo Junior College; Calvin E. Piston, John Brown University; John Putz, Alma College; Michael Schneider, Belleville Area College; Walter S. Sizer, Moorhead State University; Alan Stickney, Wittenberg University; Joseph F. Stokes, Western Kentucky University; Norman Sweet, State University College; Bill W. Vannatta, Temple Junior College; Robert P. Webber, Longwood College; Joseph E. Wiest, West Virginia Wesleyan College

ENGINEERING PRESS, INC. SAN JOSE, CALIFORNIA

Donald G. Newnan, Ph.D.
EXAM FILE Series Editor

© Copyright 1986, Engineering Press, Inc.
All rights reserved. Reproduction or translation of any part of this work beyond that permitted by section 107 or 108 of the 1976 United States copyright Act without the permission of the copyright owner is unlawful.

Printed in the United States of America

5 4 3 2 1

Library of Congress Cataloging-in-Publication Data

Calculus II exam file

(Exam file series)
1. Calculus--Problems, exercises, etc. I. Arterburn, D. R. (David.), II. Bompart, Bill.
III. Title: Calculus 2 exam file. IV. Series.
QA303.C165 1986 515'.076 86-13531
ISBN 0-910554-62-5

Engineering Press, Inc. P.O. Box 1 San Jose, California 95103-0001

Contents

Foreword

It is common practice on college campuses for student organizations to maintain files of individual professors' past exams and homework assignments. These files have helped to improve the grades of many students. For this book we have solicited exam problems from college professors all over the country, representing different approaches to the calculus topics. We have not attempted to "homogenize" the problems, preferring to leave the individual flavors intact. Each solution has been prepared by the professor who wrote the examination problem.

This volume is the second of three covering a standard college calculus course, approximately one semester of the usual three semester series. The suggested way to use this book is to choose a problem from the area of interest and work it yourself before looking at the professor's handwritten solution. We have attempted to eliminate errors, but if you should discover any, we would appreciate a note sent to Engineering Press.

We hope this book will improve your test scores, and that you will try the other volumes of calculus problems.

D. R. Arterburn
Editor

1 TRANSCENDENTAL FUNCTIONS

NATURAL LOGARITHM FUNCTION

1-1

Use logarithmic differentiation to find dy/dx if $y = x^{e^x}$.

****

$y = x^{e^x} \Rightarrow \ln y = \ln x^{e^x}$

$\ln y = (e^x)(\ln x)$

differentiate implicitly:

$\frac{dy/dx}{y} = e^x(1/x) + (\ln x)(e^x)$

solve for dy/dx: $dy/dx = y\left(\frac{e^x}{x} + (\ln x)e^x\right)$

$dy/dx = x^{e^x}(e^x)\left(\frac{1}{x} + \ln x\right)$

1-2

Let $f(x) = (\ln x)^2$ for $x > 0$. Find each of the following:

(a) critical points, and classify into relative maximums and minimums.

(b) intervals on which $f(x)$ is increasing and decreasing.

(c) inflection points.

(d) intervals on which $f(x)$ is concave up and concave down.

(c) Sketch $f(x)$.

****

$f(x) = (\ln x)^2 \quad (x>0)$

a) $f'(x) = \dfrac{2 \ln x}{x} = 0$

So, we have critical points when $\ln x = 0$ or $x = 1$, $f(x) = 0$

Note: $f'(x)$ exists for all $x > 0$, so there are <u>no</u> non-differentiable points.

b) Since $x > 0$, $f'(x) = \dfrac{2 \ln x}{x}$ is positive precisely when $\ln x > 0$ or $x > 1$

Similarly $f'(x)$ is negative when $x < 1$

So: $f(x)$ decreases for $0 < x < 1$

$f(x)$ increases for $x > 1$

(c) $f''(x) = 2 \cdot \dfrac{x \cdot 1/x - (\ln x) \cdot 1}{x^2} = 2 \cdot \dfrac{1 - \ln x}{x^2}$

$f''(x) = 0$ when $\ln x = 1$ or $x = e$

$f''(x) < 0$ when $\ln x > 1$ or $x > e$

$f''(x) > 0$ when $\ln x < 1$ or $x < e$

So, we have an inflection point at $x = e$, $f(x) = 1$

d) $f(x)$ is $\begin{cases} \text{concave up for } x > e \\ \text{concave down for } x < e \end{cases}$

Note: $\lim_{x \to \infty} (\ln x) = +\infty$, so $\lim_{x \to \infty} f(x) = +\infty$

$\lim_{x \to 0} (\ln x) = -\infty$, so $\lim_{x \to 0} f(x) = +\infty$

e) Sketch:

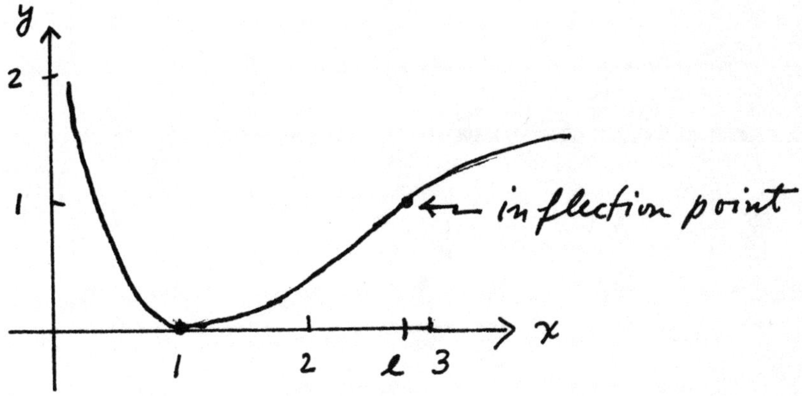

1-3

Find y' if $y = \ln \sqrt{\dfrac{4x-7}{x^2+2x}}$.

****

First, $y = \ln\left(\dfrac{4x-7}{x^2+2x}\right)^{1/2} = \dfrac{1}{2}\ln\left(\dfrac{4x-7}{x^2+2x}\right) = \dfrac{1}{2}\left(\ln(4x-7) - \ln(x^2+2x)\right)$.

Then $y' = \dfrac{1}{2}\left(\dfrac{1}{4x-7}\cdot 4 - \dfrac{1}{x^2+2x}\cdot(2x+2)\right) = \dfrac{2}{4x-7} - \dfrac{x+1}{x^2+2x}$.

1-4

If $y = \dfrac{(x+3)(x^2+1)^3(x+1)^2}{(x^2+10)^{\frac{1}{2}}}$, find y' by logarithmic differentiation.

****

$$\ln y = \ln(x+3) + 3\ln(x^2+1) + 2\ln(x+1) - \tfrac{1}{2}\ln(x^2+10)$$

$$\frac{1}{y}\cdot y' = \frac{1}{x+3} + \frac{3\cdot 2x}{x^2+1} + \frac{2}{x+1} - \frac{1}{2}\,\frac{2x}{x^2+10}$$

$$y' = \left(\frac{1}{x+3} + \frac{6x}{x^2+1} + \frac{2}{x+1} - \frac{x}{x^2+10}\right)\cdot y$$

$$y' = \left(\frac{1}{x+3} + \frac{6x}{x^2+1} + \frac{2}{x+1} - \frac{x}{x^2+10}\right)\cdot \frac{(x+3)(x^2+1)^3(x+1)^2}{(x^2+10)^{\frac{1}{2}}}.$$

1-5

Find $\dfrac{dy}{dx}$ if $y = \ln\left(\dfrac{\tan x}{x^2+1}\right)$

****

$$\frac{dy}{dx} = \frac{1}{\frac{\tan x}{x^2+1}}\cdot\left[\frac{(\sec^2 x)(x^2+1) - 2x\tan x}{(x^2+1)^2}\right]$$

$$= \frac{(x^2+1)}{\tan x}\left[\frac{(x^2+1)(\sec^2 x) - 2x\tan x}{(x^2+1)^2}\right]$$

$$= \frac{(x^2+1)\sec^2 x - 2x\tan x}{(x^2+1)\tan x}$$

$$= \frac{\sec^2 x}{\tan x} - \frac{2x}{x^2+1}.$$

$$= \frac{d}{dx}(\ln\tan x) - \frac{d}{dx}\left[\ln(x^2+1)\right]$$

1-6

Prove that $\log_{mn} x = \dfrac{\log_n x}{1 + \log_n m}$

****

using $y = \log_a x \Leftrightarrow x = a^y$

then $y = \log_{mn} x \Leftrightarrow (mn)^y = x$

that is $m^y n^y = x$

take $\log_n$ of both sides

$$\log_n m^y n^y = \log_n x$$

using $\log AB = \log A + \log B$

$$\log_n m^y + \log_n n^y = \log_n x$$

using $\log A^r = r \log A$

$$y \log_n m + y \log_n n = \log_n x$$

but $\log_n n = 1$ so $y \log_n m + y = \log_n x$

$$y(\log_n m + 1) = \log_n x$$

hence $y = \dfrac{\log_n x}{1 + \log_n m}$

That is $\log_{mn} x = \dfrac{\log_n x}{1 + \log_n m}$

NATURAL EXPONENTIAL FUNCTION

1-7

Briefly state the meaning of the term 'natural exponential function' and the term 'e'. Also, find :

$$y' \quad \text{if} \quad y = e^{\sqrt{x^3 + 1}}$$

****

The natural exponential function, denoted by 'exp', is the inverse of the natural logarithmic function. The letter 'e' denotes the positive real number such that :

$$\ln e = 1 \iff e = \exp 1$$

By the theorem for exponential function, we have:

if $u = g(x)$ and g is differentiable, then :

$$D_x e^u = e^u D_x u \qquad \ldots\ldots (1)$$

If we let $u = \sqrt{x^3 + 1}$, then by using Eq. (1) we write the given function as

$$D_x e^{\sqrt{x^3+1}} = e^{\sqrt{x^3+1}} D_x \sqrt{x^3+1}$$

$$= e^{\sqrt{x^3+1}} D_x (x^3+1)^{1/2}$$

$$= e^{\sqrt{x^3+1}} \left(\tfrac{1}{2}\right)(x^3+1)^{-1/2}(3x^2)$$

$$= e^{\sqrt{x^3+1}} \frac{3x^2}{2\sqrt{x^3+1}}$$

$$= \frac{3x^2 e^{\sqrt{x^3+1}}}{2\sqrt{x^3+1}} \qquad \underline{\text{Ans.}}$$

1-8

Let R be the region bounded by $y = \ln x$, $y = 0$, and $x = e$. Find the area of R integrating with respect to y.

****

To integrate with respect to y, we need to solve the equations for x in terms of y:

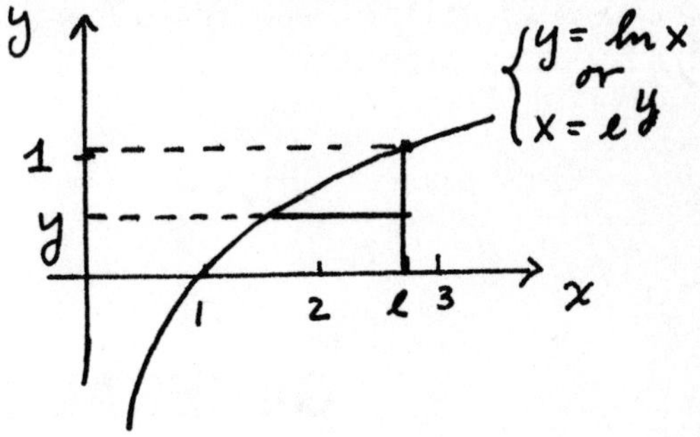

$$x = e^y, \quad y = 0, \quad x = e$$

the area of R is given by:

$$\int_0^1 (e - e^y)\,dy = \left(ey - e^y\right]_0^1 = (e\cdot 1 - e^1) - (e\cdot 0 - e^0)$$

$$= e^0 = 1$$

1-9

Find $\frac{dy}{dx}$ if $y = e^{xy} + e^{99}$

****

$$\frac{dy}{dx} = e^{xy}\left(x\frac{dy}{dx} + y\right) + 0$$

Note: e^{99} is a constant $\Rightarrow \frac{d}{dx}e^{99} = 0$.

$$\frac{dy}{dx} = xe^{xy}\frac{dy}{dx} + ye^{xy}$$

$$\Rightarrow \frac{dy}{dx}\left(1 - xe^{xy}\right) = ye^{xy}$$

$$\Rightarrow \frac{dy}{dx} = \frac{ye^{xy}}{1 - xe^{xy}}$$

1-10

Find y' if $y = xe^{(x^2+7)}$.

****

Use the product rule: $y' = x \cdot e^{x^2+7} \cdot 2x + 1 \cdot e^{x^2+7}$

$= 2x^2 e^{x^2+7} + e^{x^2+7}$.

GENERAL EXPONENTIAL AND LOGARITHMIC FUNCTIONS

1-11

Find y' if

$$y = (\sqrt{x})^x, \quad x > 0$$

****

Using logarithmic differentiation, take the natural logarithm on both sides of the equation

$$\ln(y) = \ln\left[(\sqrt{x})^x\right] = x \cdot \ln(\sqrt{x}) = \tfrac{1}{2} x \ln(x)$$

Then differentiate on both sides with respect to x.

$$\frac{1}{y} \cdot y' = \tfrac{1}{2}\left[x \cdot \frac{1}{x} + \ln(x)\right] = \tfrac{1}{2}\left[1 + \ln(x)\right]$$

$$y' = y \cdot \tfrac{1}{2}\left[1 + \ln(x)\right]$$

$$y' = \frac{(\sqrt{x})^x}{2} \cdot \left[1 + \ln(x)\right]$$

1-12

Mr. Spock has accidentally injected himself with the dangerous drug cordrazine. He quickly calculates that the concentration (y), in parts per million, of the drug in his blood t minutes after the injection is given by:

$$y = e^{-t} - e^{-2t}$$

a) At what time will the concentration reach its maximum value?

b) What will the maximum concentration be?

****

a) To maximize the concentration, solve $\frac{dy}{dt} = 0$

Now $\frac{dy}{dt} = -e^{-t} + 2e^{-2t} = e^{-t}(2e^{-t} - 1)$

So $\frac{dy}{dt} = 0 \Rightarrow e^{-t} = 0$ or $2e^{-t} - 1 = 0$

$e^{-t} = 0$ has no solution

$2e^{-t} - 1 = 0 \Rightarrow e^{-t} = 1/2 \Rightarrow -t = \ln(1/2) \Rightarrow t = \ln(2)$

Since $\frac{d^2y}{dt^2} = e^{-t} - 4e^{-2t}$ when $t = \ln(2)$,

$\frac{d^2y}{dt^2} = \frac{1}{2} - 1$ which is < 0. So by the second derivative test, $t = \ln(2)$ gives a maximum concentration.

b) Maximum concentration $= e^{-\ln 2} - e^{-2\ln 2}$

$= 1/2 - 1/4$

$= 1/4$ part per million.

1-13

Find $\frac{dy}{dx}$ for each of the following:

(a) $y = 3^{x^2} + e^{\pi} + (1 + x^2)^{\sqrt{2}}$

(b) $y = (1 + x^2)^{x^3} + \sin x$

****

(a) $\frac{dy}{dx} = 3^{x^2}(\ln 3)(2x) + 0 + \sqrt{2}(1+x^2)^{(\sqrt{2}-1)}(2x)$

(b) Use logarithmic differentiation on $w = (1+x^2)^{x^3}$

$$\ln w = x^3 \ln(1+x^2)$$

$$\frac{1}{w}\frac{dw}{dx} = 3x^2 \ln(1+x^2) + x^3\left(\frac{1}{1+x^2}\right)(2x)$$

$$\frac{dw}{dx} = w\left[3x^2 \ln(1+x^2) + \frac{2x^4}{1+x^2}\right]$$

$$= (1+x^2)^{x^3}\left[3x^2 \ln(1+x^2) + \frac{2x^4}{1+x^2}\right]$$

$$\therefore \frac{dy}{dx} = (1+x^2)^{x^3}\left[3x^2 \ln(1+x^2) + \frac{2x^4}{1+x^2}\right] + \cos x$$

1-14

Find the derivative y' for (a) $y = 3^x$ and (b) $y = x^x$.

****

(a) $y = 3^x \Rightarrow \ln y = \ln 3^x \Rightarrow \ln y = x \cdot \ln 3$. Differentiating with respect to x, $\frac{1}{y} \cdot y' = \ln 3$ or $y' = (\ln 3) \cdot y = (\ln 3) \cdot 3^x$.

(b) $y = x^x \Rightarrow \ln y = x \ln x \Rightarrow \frac{1}{y} \cdot y' = x \cdot \frac{1}{x} + 1 \cdot \ln x \Rightarrow$

$y' = (1 + \ln x) \cdot y = (1 + \ln x) \cdot x^x$.

1-15

Find $\frac{dy}{dx}$ if $y = (\ln x)^{\tan x}$

****

Taking the natural logarithm of both sides,

$$\ln y = \ln (\ln x)^{\tan x} = (\tan x) \ln(\ln x)$$

Differentiating implicitly,

$$\frac{1}{y} \frac{dy}{dx} = (\tan x)\left(\frac{1}{\ln x}\right)\left(\frac{1}{x}\right) + \ln(\ln x) \sec^2 x$$

$$\Rightarrow \frac{dy}{dx} = y\left[\frac{\tan x}{x \ln x} + \ln(\ln x) \sec^2 x\right]$$

$$\Rightarrow \frac{dy}{dx} = (\ln x)^{\tan x}\left[\frac{\tan x}{x \ln x} + \ln(\ln x) \sec^2 x\right]$$

1-16

Find the derivative of the function $y = x^{\sin x}$ with and without using logarithmic differentiation.

****

By normal differentiation, we have

$$y = x^{\sin x}$$

$$f(x) = e^{\sin x . \ln x}$$

$$f'(x) = e^{\sin x . \ln x}\left(\sin x . \frac{1}{x} + \ln x . \cos x\right)$$

[By Product Rule]

$$= x^{\sin x}\left(\frac{1}{x} . \sin x + \cos x . \ln x\right)$$

Ans.

By logarithmic differentiation, we write

$$y = x^{\sin x}$$

or $$\ln y = \ln x^{\sin x}$$

$$\frac{1}{y} . \frac{dy}{dx} = \frac{d}{dx}\left[\sin x . \ln x\right]$$

$$= \sin x . \frac{1}{x} + \ln x . \cos x \quad \left[\text{By Product Rule}\right]$$

Hence,

$$\frac{dy}{dx} = y\left(\frac{1}{x} . \sin x + \cos x . \ln x\right)$$

$$= x^{\sin x}\left(\frac{1}{x} . \sin x + \cos x . \ln x\right)$$

Ans.

Note that both the methods give the same answer.

1-17

Give the value of each of the following:

a) $\log_{32} 8$

b) $25^{\log_5 10}$

****

a) let $x = \log_{32} 8$

then $32^x = 8$; writing both 32 and 8 as powers of 2, we have

$$2^{5x} = 2^3, \text{ thus } 5x = 3$$
$$x = 3/5$$

therefore $\boxed{\log_{32} 8 = 3/5}$

b) let $x = 25^{\log_5 10}$

We want the form $A^{\log_A B} = B$, so we write 25 as 5^2, giving

$$x = 5^{2\log_5 10}$$
$$= 5^{\log_5 10^2}$$
$$= 5^{\log_5 100} \boxed{= 100}$$

HYPERBOLIC FUNCTIONS

1-18

$\lim\limits_{x\to+\infty} \tanh x =$ (a) 0 (b) 1 (c) $+\infty$ (d) $-\infty$ (e) -1.

****

$$\lim_{x\to+\infty} \tanh x = \lim_{x\to+\infty} \frac{e^x - e^{-x}}{e^x + e^{-x}} = \lim_{x\to+\infty} \frac{1 - \frac{1}{e^{2x}}}{1 + \frac{1}{e^{2x}}} = \frac{1-0}{1+0} = 1.$$

1-19

If sin ϕ = tanh x find the value of (a) tan ϕ (b) sec ϕ

****

If $\sin\phi = \tanh x$

then we can use a triangle to determine $\tan\phi$ and $\sec\phi$

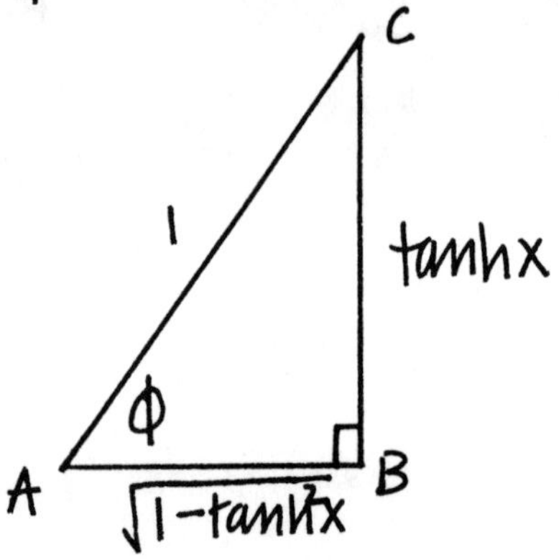

since $\sin\phi = \frac{OPP}{HYP}$

so the adjacent side AB has length $\sqrt{1-\tanh^2 x}$ by Pythagoras' theorem

but $\tanh^2 x + \operatorname{Sech}^2 x = 1$

so AB has length sech x

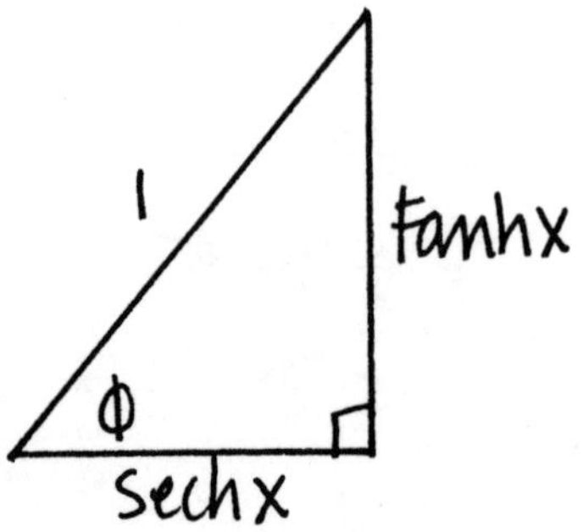

now $\tan\phi = \frac{\tanh x}{\operatorname{sech} x}$

$= \frac{\sinh x}{\cancel{\cosh} x} \cdot \cancel{\cosh} x$

and $\sec\phi = \frac{1}{\frac{\operatorname{Sech} x}{1}}$ since $\sec\phi$ is $\frac{1}{\cos\phi}$

$= 1/\frac{1}{\cosh x} = \cosh x$

<u>hence $\tan\phi = \sinh x$ and $\sec\phi = \cosh x$</u>

1-20

Show that $\sinh[\ln(x)] = \dfrac{x^2 - 1}{2x}$

****

Since $\sinh(u) = \dfrac{e^{u} - e^{-u}}{2}$,

$$\sinh[\ln(x)] = \frac{e^{\ln(x)} - e^{-\ln(x)}}{2} = \frac{e^{x} - e^{\ln(1/x)}}{2}$$

$$= \frac{x - \frac{1}{x}}{2} = \frac{\frac{x^2-1}{x}}{2} = \frac{x^2-1}{2x}$$

1-21

Find the exact coordinates of the two points of inflection of the graph of y = sech x.

****

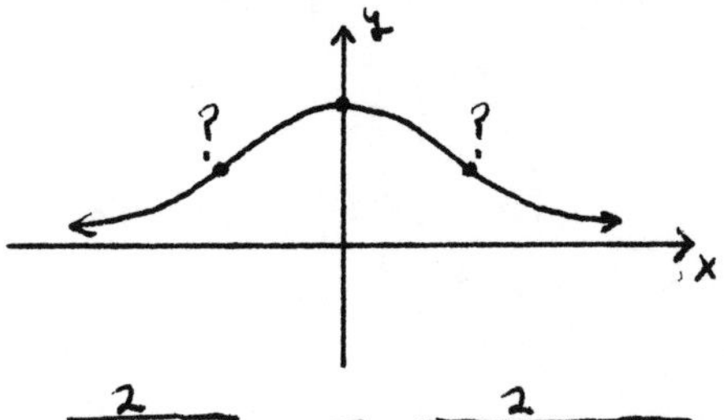

Differentiating twice, we get $y'' = -\operatorname{sech}^3 x + \operatorname{sech} x \tanh^2 x$. So $y'' = 0 \Rightarrow \operatorname{sech} x \tanh^2 x = \operatorname{sech}^3 x \Rightarrow \tanh^2 x = \operatorname{sech}^2 x \Rightarrow \sinh^2 x = 1 \Rightarrow \sinh x = \pm 1$. For $\sinh x = 1$, $x = \sinh^{-1} 1 = \ln(1+\sqrt{2})$,

$y = \dfrac{2}{e^x + e^{-x}} = \dfrac{2}{1+\sqrt{2} + \frac{1}{1+\sqrt{2}}} =$ (via simplification) $\dfrac{\sqrt{2}}{2}$. Therefore, the points of inflection are $\left(\pm \ln(1+\sqrt{2}), \dfrac{\sqrt{2}}{2}\right)$.

1-22

Evaluate $\int \cosh(\ln x)dx$.

****

$$\int \cosh(\ln x)dx = \int \frac{e^{\ln x} + e^{-\ln x}}{2} dx$$

$$= \int \frac{x + \frac{1}{x}}{2} dx \quad = \frac{x^2}{4} + \frac{1}{2}\ln x + c$$

INVERSE OF A FUNCTION

1-23

Find the inverse of $f(x) = x^2-2x + 3 \quad (x \geq 1)$

****

The restriction $x \geq 1$ assures us that the function is one-to-one. To find the inverse proceed as follows:

Step 1: Write $y = x^2 - 2x + 3$, $x \geq 1$

Step 2: Interchange x and y in step one to get $x = y^2 - 2y + 3$, $y \geq 1$.

Step 3: Solve for y in step two. Rewrite as $y^2 - 2y = x - 3$, $y \geq 1$. Complete the square: $y^2 - 2y + 1 = x - 2$, $y \geq 1$. Factor: $(y-1)^2 = x - 2$, $y \geq 1$. Take roots: $y - 1 = \pm\sqrt{x-2}$, $y \geq 1$. Add one: $y = 1 \pm \sqrt{x-2}$, $y \geq 1$. To decide the proper root note the restriction $y \geq 1$ means $y - 1 \geq 0$. Hence the left side in the next to last equation is positive and we must choose $+\sqrt{x-2}$. Formally we have

$f^{-1}(x) = 1 + \sqrt{x-2}$, $y \geq 1$. As a check note that $f(f^{-1}(x)) = f(1 + \sqrt{x-2}) = (1 + \sqrt{x-2})^2 - 2(1 + \sqrt{x-2}) + 3 =$ $1 + 2\sqrt{x-2} + (x-2) - 2 - 2\sqrt{x-2} + 3 = 1 + x - 4 + 3 = x$.

Similarly, $f^{-1}(f(x)) = x$ also.

1-24

Find the inverse of the function $f(x) = \sqrt{x-5}$. Be sure to write the domain of the inverse function.

****

LET $g(x)$ BE THE INVERSE OF $f(x)$.
THEN $f(g(x)) = x$

$$\sqrt{g(x)-5} = x$$
$$g(x)-5 = x^2$$
$$g(x) = x^2+5$$

SINCE THE DOMAIN OF AN INVERSE IS THE RANGE OF THE FUNCTION, THE DOMAIN OF g IS THE SET OF ALL x SUCH THAT $x \geq 0$. EXPRESSED AS AN INTERVAL, THE DOMAIN IS $[0, \infty)$.

1-25

Suppose $G(x) = \int_1^x \sqrt{t^2 + 3}\, dt$. Does the function G have an inverse function? Justify your answer.

****

By the Fundamental Theorem, $G'(x) = \sqrt{x^2+3}$.

So, $G'(x) > 0$ for all x.

$\therefore$ G is an increasing function and G must have an inverse function.

1-26

If $f(x) = x^2 - x - 6$, $x \geq 1$, find $(f^{-1})'(6)$ by two methods.

****

a.) If $y = f(x) = x^2 - x - 6$, $x \geq 1$ defines f,

$x = y^2 - y - 6$, $y \geq 1$ defines f^{-1}.

$0 = y^2 - y + (-6 - x)$ is a quadratic equation which has

$$y = \frac{1 \pm \sqrt{25 + 4x}}{2} \text{ as solutions.}$$

Since $y \geq 1$,

$$y = f^{-1}(x) = \frac{1 + \sqrt{25 + 4x}}{2}$$

and $(f^{-1})'(x) = \frac{1}{2} \cdot \frac{1}{2}(25 + 4x)^{-1/2}(4) = \dfrac{1}{\sqrt{25 + 4x}}$

$$\therefore (f^{-1})'(6) = \frac{1}{\sqrt{25 + 4(6)}} = \frac{1}{\sqrt{49}} = \frac{1}{7}$$

b) By the chain rule $\frac{dy}{dx} \cdot \frac{dx}{dy} = 1$ and

$$\frac{dx}{dy} = \frac{1}{\frac{dy}{dx}} \quad \text{or} \quad D_y x = \frac{1}{D_x y}$$

where $D_x y$ is the derivative of f and

$D_y x$ is the derivative of f^{-1}.

$y' = 2x - 1$, so $D_y x = \dfrac{1}{2x - 1}$. Now find

what x in the domain of f is paired with 6, which is in the domain of f^{-1}. To do this, set $x^2 - x - 6 = 6$ and solve.

$$x^2 - x - 6 = 6$$
$$x^2 - x - 12 = 0$$

which has $x = 4$ and $x = -3$ as solutions. Since $x \geq 1$, $x = 4$ is the only solution.

$$\therefore D_y x = \frac{1}{2X-1}\Big|_4 = \frac{1}{2(4)-1} = 1/7$$

This checks with answer in part (a).

1-27

Sketch the graph of f for $f(x) = \sqrt[3]{x}$ and determine if f^{-1} exists. If so, find a formula for $y = f^{-1}(x)$ and also sketch the graph of f^{-1}.

****

A set of coordinate values for $f(x) = \sqrt[3]{x}$ is shown below, from which the accompanying graph is drawn.

x	0	0.5	1	2	3	4
y	0	0.79	1	1.26	1.44	1.58

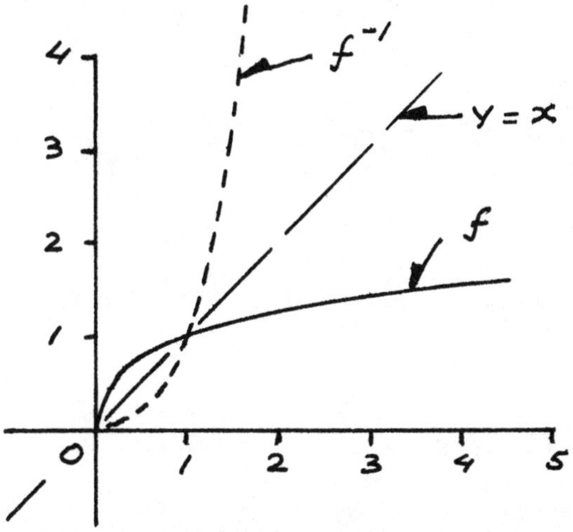

We note from the graph that f^{-1} exists because no horizontal line intersects the graph of $f(x) = \sqrt[3]{x}$ more than once, i.e. the function is single valued.

We are given $y = \sqrt[3]{x}$

Hence, for inverse, we have:

$$x = \sqrt[3]{y}$$

$$\text{OR} \quad x^3 = y$$

$$= f^{-1}(x)$$

The inverse graph of $y = \sqrt[3]{x}$, i.e. $x = \sqrt[3]{y}$ is shown as a dotted line.

1-28

Find $f^{-1}(x)$ for $f(x) = \sqrt{e^x + 2}$.

****

$f(x) = (e^x+2)^{1/2}$ inverse $x = (e^{f^{-1}(x)}+2)^{1/2}$

$\longrightarrow x^2 = e^{f^{-1}(x)}+2 \rightarrow x^2-2 = e^{f^{-1}(x)}$

$\longrightarrow f^{-1}(x) = \ln(x^2-2)$

INVERSE TRIGONOMETRIC FUNCTIONS

1-29

With derivatives prove that $\arcsin \frac{1}{x} = \operatorname{arccsc}(x)$ for all $x \geq 1$.

****

Let $f(x) = \arcsin x^{-1}$ and $g(x) = \operatorname{arccsc} x$ for all $x \in [1,\infty)$. Now we have

$$f'(x) = \frac{-\frac{1}{x^2}}{\sqrt{1-\frac{1}{x^2}}} = \frac{-\frac{1}{x^2}}{\sqrt{\frac{x^2-1}{x^2}}} = \frac{-\frac{1}{x^2}}{\frac{\sqrt{x^2-1}}{x}} = \frac{-1}{x\sqrt{x^2-1}}$$

and $g'(x) = \dfrac{-1}{x\sqrt{x^2-1}}$. Since $f'(x) = g'(x)$ we may conclude that $f(x)$ and $g(x)$ differ at most by a constant. So $f(x) - g(x) = K$ and we must find K. Let $x = 1$ then $f(1) - g(1) = \arcsin 1 - \operatorname{arccsc} 1 = \frac{\pi}{2} - \frac{\pi}{2} = 0$. Since $K = 0$ we have that $f(x) = g(x)$ i.e $\arcsin \frac{1}{x} = \operatorname{arccsc} x$.

1-30

Find an expression for cos[arc tan(sin(arc cot x))] in terms of x

****

$\cos[\operatorname{arctan}(\sin(\operatorname{arccot} x))]$

let $y = \operatorname{arccot} x$ then $\cot y = x$

so we can complete the triangle

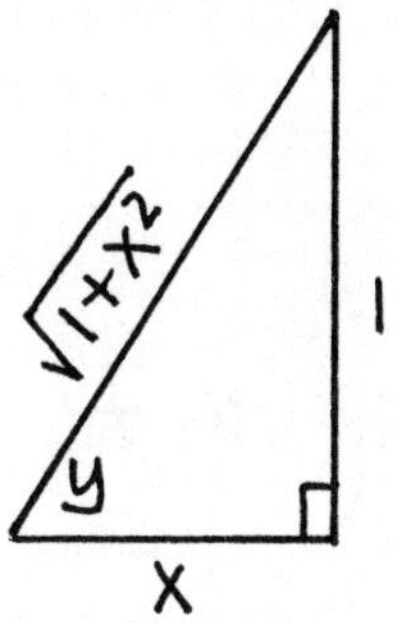

now $\sin(\operatorname{arccot} x) = \sin y$

from the triangle

$$\sin y = \frac{1}{\sqrt{1+x^2}}$$

so we now need to find $\cos\left[\operatorname{arctan}\frac{1}{\sqrt{1+x^2}}\right]$

let $z = \operatorname{arctan}\frac{1}{\sqrt{1+x^2}}$, so $\tan z = \frac{1}{\sqrt{1+x^2}}$

so, again we can complete the triangle

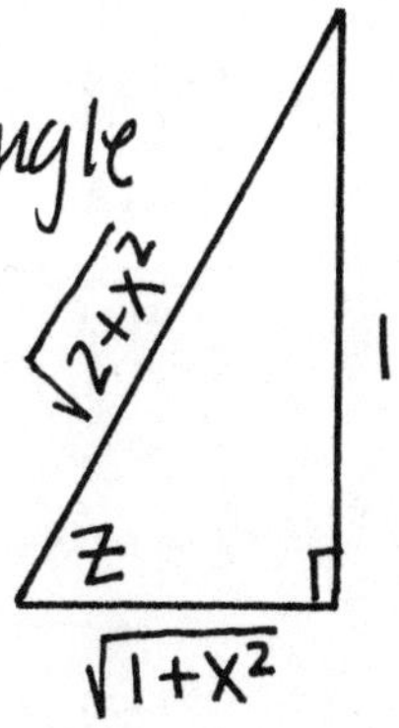

finally $\cos z = \dfrac{\sqrt{1+x^2}}{\sqrt{2+x^2}}$

Hence $\cos[\operatorname{arctan}(\sin(\operatorname{arccot} x))] = \sqrt{\dfrac{1+x^2}{2+x^2}}$

1-31

For $y = \dfrac{\text{Arcsin } x}{\sqrt{1 - x^2}}$, $|x| < 1$, find y". Simplify your answer.

****

$$y' = \frac{(1-x^2)^{1/2}\left(\frac{1}{(1-x^2)^{1/2}}\right) - (\text{Arcsin } x)(\tfrac{1}{2})(1-x^2)^{-1/2}(-2x)}{\left((1-x^2)^{1/2}\right)^2}$$

$$= \frac{1 + \dfrac{x \text{ Arcsin } x}{(1-x^2)^{1/2}}}{1-x^2} = \frac{(1-x^2)^{1/2} + x \text{ Arcsin } x}{(1-x^2)^{3/2}}$$

$$y'' = \frac{(1-x^2)^{3/2}\left[(1/2)(1-x^2)^{-1/2}(-2x) + x(1-x^2)^{-1/2} + \text{Arcsin } x\right] - \left[(1-x^2)^{1/2} + x \text{ Arcsin } x\right](3/2)(1-x^2)^{1/2}(-2x)}{(1-x^2)^3}$$

$$= \frac{(1-x^2)^{3/2} \text{ Arcsin } x + 3x(1-x^2)^{1/2}\left((1-x^2)^{1/2} + x \text{ Arcsin } x\right)}{(1-x^2)^3}$$

$$= \frac{(1-x^2)^{1/2}\left[(1-x^2)\text{ Arcsin } x + 3x\left((1-x^2)^{1/2} + x \text{ Arcsin } x\right)\right]}{(1-x^2)^3}$$

$$= \frac{(1+2x^2)(\text{Arcsin } x) + 3x(1-x^2)^{1/2}}{(1-x^2)^{5/2}}$$

1-32

Find the exact value of: $\sin \left[2 \cos^{-1} \left(- \frac{7}{8} \right) \right]$.

****

We wish to use some known geometric identities for solving a problem of this kind, and hence we write:

Let $\quad t = \cos^{-1}\left(-\frac{7}{8}\right) \quad \cdots\cdots(1)$

Therefore, our problem statement now reduces to finding the exact value of:

$$\sin 2t$$

This suggests that we can use the identity: $\quad \sin 2t = 2 \sin t \cos t \quad \cdots\cdots(2)$

By rewriting Eq.(1) we have

$$\cos t = -\frac{7}{8}, \qquad \text{for } 0 < t < \pi$$

Now we use the identity

$$\sin^2 t + \cos^2 t = 1 \qquad \cdots\cdots(3)$$

and note that $\sin t > 0$ because $0 < t < \pi$

By rewriting Eq.(3) we have

$$\sin t = \sqrt{1 - \cos^2 t}$$

$$= \sqrt{1 - \left(-\frac{7}{8}\right)^2} = 0.484$$

Substituting the calculated values of $\sin t$ and $\cos t$ in Eq.(2) we get

$$\sin 2t = 2\,(0.484)\left(-\frac{7}{8}\right) = -0.847$$

Hence, $\sin\left[2 \cos^{-1}\left(-\frac{7}{8}\right)\right] = -0.847$ **Ans.**

NOTE: This problem has been solved for t in the interval $0 < t < \pi$. Similarly, if we were to consider the interval $-\pi < t < 0$, we would find that:

$$\sin\left[2 \cos^{-1}\left(-\frac{7}{8}\right)\right] = 0.847.$$

1-33

Find the following values exactly:

(a) arcsin(sin(9π/7))

(b) tan(arccos(-1/3))

****

(a) Find an angle θ with $-\frac{\pi}{2} \leq \theta \leq \frac{\pi}{2}$ and $\sin\theta = \sin\frac{9\pi}{7}$.

In the diagram, P and Q have the same y-coordinate

$\therefore \quad \theta = -\frac{2\pi}{7}$

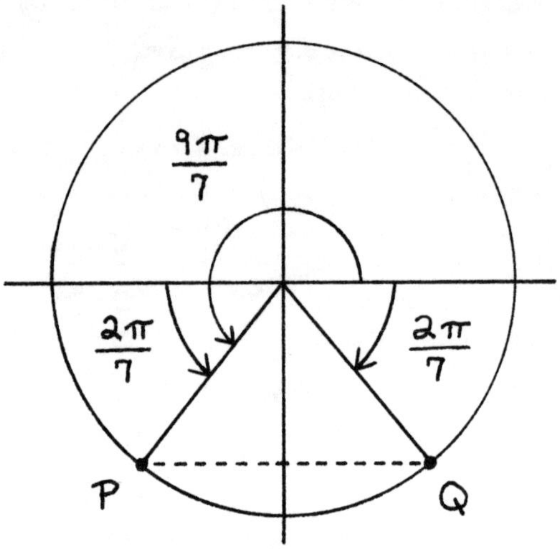

(b) Let $\theta = \arccos\left(\frac{-1}{3}\right)$

Then θ is shown in the diagram.

$\therefore \quad \tan\theta = -\sqrt{8}$

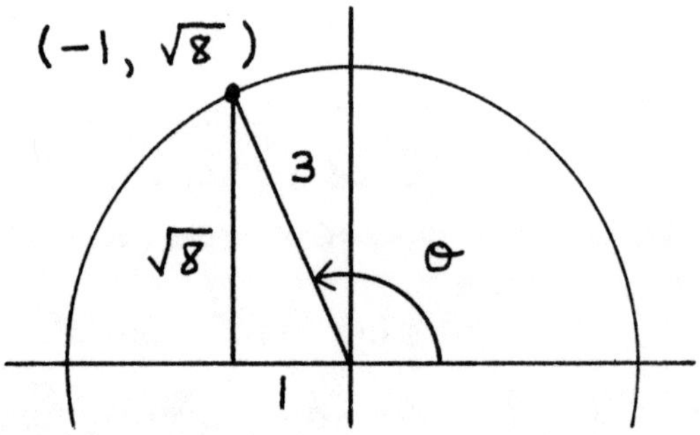

1-34

Use differentials to find an approximation for $\cos^{-1}(0.13)$ to 3 decimal places.

****

Let $f(x) = \cos^{-1}(x)$ then $f'(x) = -\frac{1}{\sqrt{1-x^2}}$.

$\therefore f(x+\Delta x) = \cos^{-1}(0.13)$

$= \cos^{-1}(0+.13)$ where $x=0$ and $\Delta x = .13$

But $f(x+\Delta x) \approx f(x) + f^{-1}(x)\Delta x$ so

$$\cos^{-1}(0.13) \approx f(0) + f^{-1}(0)\cdot(.13)$$

$$\approx \cos^{-1}(0) + \left(-\frac{1}{\sqrt{1-0}}\right)(.13)$$

$$\approx \frac{\pi}{2} - .13$$

$$\approx 1.5708 - .13$$

$$\approx 1.441$$

Checking with a calculator, $\cos^{-1}(0.13) \approx 1.440$.

1-35

If $x = \frac{1}{3}\cos\theta$, what is $\tan^2\theta$ in terms of x ?

****

If $x = \frac{1}{3}\cos\theta$, then $3x = \cos\theta$ and $\theta = \cos^{-1}3x$

this suggests the right triangle shown and labeled as in the figure.

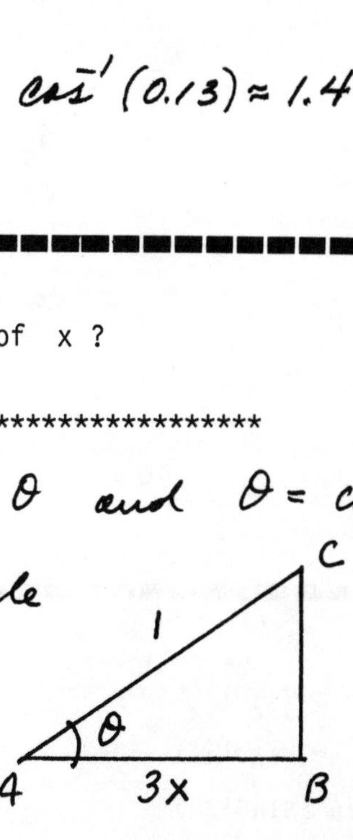

By Pythagoras,

$BC^2 = 1 - 9x^2$, so $\tan^2\theta = \frac{BC^2}{AB^2} = \frac{1-9x^2}{9x^2}$

1-36

Find the value of $\displaystyle\int_0^{\frac{1}{\sqrt{3}}} \frac{dx}{1+3x^2}$

****

$$\int_0^{\frac{1}{\sqrt{3}}} \frac{dx}{1+3x^2} = \int_0^{\frac{1}{\sqrt{3}}} \frac{dx}{1+(\sqrt{3}x)^2} .$$

Let $u = \sqrt{3}\,x$
$du = \sqrt{3}\,dx$

$$= \int_{x=0}^{x=\frac{1}{\sqrt{3}}} \frac{du/\sqrt{3}}{1+u^2} = \frac{1}{\sqrt{3}} \arctan u \Big]_{x=0}^{x=\frac{1}{\sqrt{3}}}$$

$$= \frac{1}{\sqrt{3}} \arctan(\sqrt{3}x) \Big]_0^{1/\sqrt{3}}$$

$$= \frac{1}{\sqrt{3}} \arctan 1 - \frac{1}{\sqrt{3}} \arctan 0$$

$$= \frac{1}{\sqrt{3}}\left(\frac{\pi}{4}\right) - 0 = \frac{\pi}{4\sqrt{3}}$$

1-37

Find each of the following:

a) $\sin(\text{Sin}^{-1}3x)$

b) $\cos(\text{Sin}^{-1}3x)$

c) $\tan(\text{Sin}^{-1}3x)$

****

a) $\sin(\mathrm{Sin}^{-1} 3x) = \underline{\underline{3x}}$ $\quad (-\frac{1}{3} \le x \le \frac{1}{3})$ by definition.

b) Substitute $y = \mathrm{Sin}^{-1} 3x$, then $\sin y = 3x$, $-\frac{\pi}{2} \le y \le \frac{\pi}{2}$.

Since $\sin^2 y + \cos^2 y = 1$, $\cos^2 y = 1 - \sin^2 y$ and

$\cos y = \oplus \sqrt{1 - \sin^2 y}$. To decide the proper sign note the restriction forces us to choose y in quadrant I or IV where $\cos y$ is positive.
Hence $\cos y = \oplus \sqrt{1 - \sin^2 y}$ and by back substitution we have

$$\cos(\mathrm{Sin}^{-1} 3x) = \underline{\underline{\sqrt{1 - 9x^2}}}.$$

c) By the Ratio identity we have

$$\mathrm{Tan}(\mathrm{Sin}^{-1} 3x) = \frac{\sin(\mathrm{Sin}^{-1} 3x)}{\cos(\mathrm{Sin}^{-1} 3x)} = \frac{3x}{\sqrt{1 - 9x^2}}.$$

INVERSE HYPERBOLIC FUNCTIONS

1-38

Let argtanh represent the inverse of the hyperbolic tangent function. Derive an equation which gives argtanh x in terms of the natural logarithm of a function of x.

****

Let $y = \mathrm{argtanh}\, x = \tanh^{-1} x$. Then $\tanh y = x$,

so $\dfrac{e^y - e^{-y}}{e^y + e^{-y}} = x$. $\quad e^y - e^{-y} = x(e^y + e^{-y})$.
Multiply by e^y to get

$e^{2y} - 1 = x(e^{2y} + 1)$. Then $e^{2y}(1 - x) = 1 + x$.
$e^{2y} = \dfrac{1+x}{1-x}$. Now take the natural logarithm of both sides. $\quad 2y = \ln\left|\dfrac{1+x}{1-x}\right|$. $\quad y = \ln\sqrt{\dfrac{1+x}{1-x}}$.

$$\mathrm{Argtanh}\, x = \ln\sqrt{\frac{1+x}{1-x}} \text{ for } -1 < x < 1.$$

1-39

Find an explicit formula for $\sinh^{-1}x$.

****

$y = \sinh^{-1}x$ if and only if $x = \sinh y$,

$x = \frac{e^y - e^{-y}}{2}$. Solve for y: $e^y - e^{-y} = 2x$,

multiply by e^y: $e^{2y} - 2xe^y - 1 = 0$.

This is a quadratic in e^y, so

$$e^y = \frac{2x \pm \sqrt{4x^2+4}}{2},$$

$e^y = x \pm \sqrt{1+x^2}$. Since $e^y > 0$, take

$+$ sign, $e^y = x + \sqrt{1+x^2}$.

$$\sinh^{-1}x = y = \ln(x + \sqrt{1+x^2}).$$

1-40

Find dy/dx if $x\tanh^{-1}(y) + y/x = 0$

****

Using implicit differentiation

$$d[x\tanh^{-1}(y)] + d\left(\frac{y}{x}\right) = d(0)$$

$$x \cdot \frac{1}{1-y^2}\,dy + \tanh^{-1}(y)\,dx + \left[\frac{x\,dy - y\,dx}{x^2}\right] = 0$$

Multiplying by x^2,

$$\frac{x^3}{1-y^2}\,dy + x^2\tanh^{-1}(y)\,dx + x\,dy - y\,dx = 0$$

$$\left(\frac{x^3}{1-y^2} + x\right) dy = \left(y - x^2 \tanh^{-1}(y)\right) dx$$

$$\left(\frac{x^3 + x - xy^2}{1-y^2}\right) dy = \left(y - x^2 \tanh^{-1}(y)\right) dx$$

$$\frac{dy}{dx} = \frac{(1-y^2)\left[y - x^2 \tanh^{-1}(y)\right]}{x^3 + x - xy^2}$$

DERIVATIVES OF TRIGONOMETRIC FUNCTIONS

1-41

Differentiate each of the following:

a) $f(t) = \mathrm{Tan}^{-1} t$

b) $f(t) = \mathrm{Tan}^{-1} \sqrt{1-t}$

****

a) Substitute $y = f(t)$ then $y = \mathrm{Tan}^{-1} t$ and by def.

$\mathrm{Tan}\, y = t \quad (-\frac{\pi}{2} < y < \frac{\pi}{2})$. Now differentiate implicitly with Respect to t: $\sec^2 y \left(\frac{dy}{dt}\right) = 1$ OR $\frac{dy}{dt} = \frac{1}{\sec^2 y}$. But $\sec^2 y = 1 + \tan^2 y$. Hence we have

$$\frac{dy}{dt} = \underline{\underline{f'(t)}} = \frac{1}{1 + \mathrm{Tan}^2 y} = \frac{1}{1+t^2} \text{ by substitution.}$$

b) Substitute $y = f(t)$ and $u = \sqrt{1-t}$ then $y = \mathrm{Tan}^{-1} u$ and by the chain Rule $\frac{dy}{dt} = \frac{dy}{du} \cdot \frac{du}{dt}$. By part a)

$\frac{dy}{du} = \frac{1}{1+u^2}$. Also $\frac{du}{dt} = \frac{1}{2}(1-t)^{-1/2}(-1) = \frac{-1}{2\sqrt{1-t}}$. Hence

$$\frac{dy}{dt} = f'(t) = \frac{1}{1 + (1-t)} \cdot \frac{-1}{2\sqrt{1-t}} = \frac{-1}{(4-2t)\sqrt{1-t}}.$$

1-42

Given $f(x) = \cos^{-1}(x^2) + \sin^{-1}(x^2)$,

a) Evaluate $f(1/\sqrt{2})$

b) Find $f'(x)$. (simplify your answer)

c) Evaluate $f(.13457)$. (This should be able to be done without a calculator.

****

a) $$f(1/\sqrt{2}) = \cos^{-1}(1/\sqrt{2})^2 + \sin^{-1}(1/\sqrt{2})^2$$
$$= \cos^{-1}(1/2) + \sin^{-1}(1/2)$$
$$= \pi/3 + \pi/6$$
$$= \boxed{\pi/2}$$

b) $$f'(x) = \left[\cos^{-1}(x^2) + \sin^{-1}(x^2)\right]'$$
$$= \frac{-1}{\sqrt{1-(x^2)^2}} \cdot 2x + \frac{1}{\sqrt{1-(x^2)^2}} \cdot 2x$$
$$= \boxed{0}$$

c) Since $f'(x) = 0$ and $f'(x)$ exists for all x in $[-1, 1]$

$$f(.13457) = \boxed{\pi/2}$$

1-43

Find the derivative for $f(x) = \cos(\sin^2 x)$.

****

By the chain rule

$$f'(x) = -\sin(\sin^2 x) \frac{d}{dx}[\sin^2 x]$$

Applying the chain rule again gives

$$= -\sin(\sin^2 x) \cdot 2\sin x \frac{d}{dx}[\sin x]$$

$$= -\sin(\sin^2 x) \cdot 2\sin x \cos x$$

By a double angle identity

$$= \underline{\underline{-\sin(\sin^2 x)\sin 2x}}$$

Note that in general $\sin(\sin^2 x) \neq \sin^3 x$

1-44

Find the equation of the tangent line to the curve y = sec x - csc 2x at the point with abscissa $\pi/6$.

****

$$dy/dx = \sec x \tan x - [-2\csc 2x \cot 2x]$$

$$= \sec x \tan x + 2\csc 2x \cot 2x$$

$$dy/dx\Big|_{x=\pi/6} = \sec \pi/6 \tan \pi/6 + 2\csc \pi/3 \cot \pi/3$$

$$= (2/\sqrt{3})(\sqrt{3}/3) + 2(2/\sqrt{3})(1/\sqrt{3})$$

$$= \frac{2}{3} + \frac{4}{3} = 2$$

$$y(\pi/6) = \sec \pi/6 - \csc \pi/3 = \frac{2}{\sqrt{3}} - \frac{2}{\sqrt{3}} = 0$$

$\therefore$ The tangent line at $(\pi/6, 0)$ has equation:

$$y = 2(x - \pi/6)$$

or $y = 2x - \pi/3$

1-45

Find the derivative dy/dx (where it exists) for the following:

$$y \tan^2 x + \sec y = 1$$

****

Implicit differentiation:

$$\frac{d}{dx}(y \tan^2 x + \sec y) = \frac{d}{dx}(1)$$

$$\Leftrightarrow \frac{d}{dx}(y \tan^2 x) + \frac{d}{dx}(\sec y) = \frac{d}{dx}(1)$$

$$\frac{d}{dx}(y \tan^2 x) = y(2 \tan x \sec^2 x) + \tan^2 x \frac{dy}{dx}$$

$$\frac{d}{dx}(\sec y) = \sec y \tan y \frac{dy}{dx}$$

$$\frac{d}{dx}(1) = 0$$

Thus $2y \tan x \sec^2 x + \tan^2 x \frac{dy}{dx} + \sec y \tan y \frac{dy}{dx} = 0$

$$\Leftrightarrow \frac{dy}{dx}\left(\tan^2 x + \sec y \tan y\right) = -2y \tan x \sec^2 x$$

$$\Leftrightarrow \frac{dy}{dx} = \frac{-2y \tan x \sec^2 x}{\tan^2 x + \sec y \tan y}$$

1-46

Derive the formula for the derivative of the inverse hyperbolic sine function.

****

$y = \sinh^{-1} x \Rightarrow \sinh y = x$. Differentiating implicitly with respect to x, $\cosh y \cdot y' = 1$ so $y' = \frac{1}{\cosh y} = \frac{1}{\sqrt{\sinh^2 y + 1}} = \frac{1}{\sqrt{x^2+1}}$, ie. $D_x \sinh^{-1} x = \frac{1}{\sqrt{x^2+1}}$.

1-47

Find the derivative of $(\ln \sec x)^4$.

****

$$\frac{d}{dx}(\ln \sec x)^4 = 4(\ln \sec x)^3 \frac{1}{\sec x} \sec x \tan x$$

$$= 4 \tan x (\ln \sec x)^3$$

APPLICATIONS OF THE EXPONENTIAL AND LOGARITHMIC FUNCTIONS

1-48

Suppose pH= $-\log [H^+]$. Suppose further that for vinegar, the hydrogen ion concentration in moles per liter, is given by $[H^+]= 5.2(10^{-4})$. Find the pH of the vinegar.

****

$[H^+] = (5.2)(10^{-4})$. Hence

$$\log [H^+] = \log [(5.2)(10^{-4})]$$

$$= \log 5.2 + \log (10^{-4})$$

$$= \log 5.2 + (-4)$$

So

$$-\log [H^+] = 4 - \log 5.2$$

Therefore $pH = 4 - \log 5.2$

$$= 3.28$$

1-49

A body traveling in a medium is retarded by a force proportional to its velocity. If the velocity of the body after 4 seconds is 80 ft./sec. and the velocity after 6 seconds is 60 ft./sec., find the initial velocity of the body.

****

$$\frac{dv}{dt} = -kv$$

$$\frac{dv}{v} = -k\,dt.$$

INTEGRATE BOTH SIDES.

$$\ln v = -kt + C.$$

SO

$$v = v_0 e^{-kt}$$

AT $t=4$ $v=80$ and $t=6$; $v=60$

SO

$$80 = v_0 e^{-4k}$$

$$60 = v_0 e^{-6k}$$

THUS

$$\ln 80 = \ln v_0 - 4k$$

$$\ln 60 = \ln v_0 - 6k.$$

$$\ln 80 - \ln 60 = 2k$$

$$\ln 4/3 = 2k$$

THUS $k = \frac{1}{2} \ln 4/3$

SO

$$v = v_0 e^{-\frac{t}{2} \ln 4/3}$$

if $t = 4$ then $v = 80$ so

$$80 = v_0 e^{-2 \ln 4/3} = v_0 e^{\ln (3/4)^2} = \frac{9}{16} v_0$$

THUS $v_0 = \frac{16 \cdot 80}{9} = \frac{1280}{9}$

━━━━━━━━━━━━━━━━━━━━━━━━━━━━━━━━━━━━━**1-50**

Determine whether $f(x) = e^{\frac{1}{x}}$ has any points of inflection

****

> A function has a point of inflection at x=a if
> (i) $f''(x)=0$ at x=a
> (ii) curvature changes at x=a.

finding $f'(x)$ using the chain rule.

$$f'(x) = e^{\frac{1}{x}} \cdot \frac{-1}{x^2} = \frac{-e^{\frac{1}{x}}}{x^2}$$

finding $f''(x)$ using the quotient and chain rules.

$$f''(x) = \frac{x^2\left(+\frac{e^{\frac{1}{x}}}{x^2}\right) - (-e^{\frac{1}{x}})2x}{x^4}$$

$$= \frac{+e^{\frac{1}{x}}+2xe^{\frac{1}{x}}}{x^4} = \frac{e^{\frac{1}{x}}}{x^4}(2x+1)$$

note $e^{\frac{1}{x}}>0$ and $x^4>0$, so $f''(x)=0$ when $x=-\frac{1}{2}$

$$
\begin{array}{ll}
e^{\frac{1}{x}} & +\ +\ +\ +\ +\ +\ +\ +\ +\ +\ + \\
x^4 & +\ +\ +\ +\ +\ +\ +\ +\ +\ +\ + \\
2x+1 & -\ -\ -\ \ +\ +\ +\ +\ +\ +\ +
\end{array}
$$

⤷ $-\frac{1}{2}$ ⤵ 0 ⤴

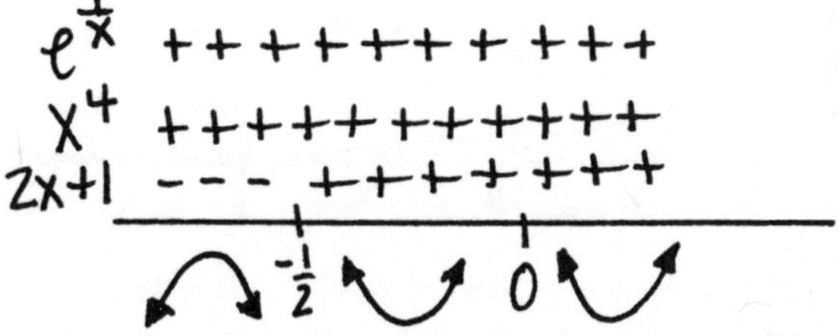

notice there is also a
discontinuity at x=0
$\lim\limits_{x\to 0^+} e^{\frac{1}{x}} = e^{\infty} = \infty$
$\lim\limits_{x\to 0^-} e^{\frac{1}{x}} = e^{-\infty} = 0$.

Hence $e^{\frac{1}{x}}$ has an inflection point at $x=-\frac{1}{2}$

But this is not a point
of inflection since
$f''(0) \neq 0$

1-51

A cost function, C, is proportional to the marginal cost function. Fixed cost is $500. When 3 units are produced, the cost is $1500. Find the function.

****

$C(x) = Ke^{\alpha x}$

$500 = Ke^{\alpha(0)}$

$500 = K$

$C(x) = 500\,e^{\alpha x}$

$1500 = 500\,e^{\alpha \cdot 3}$

$3 = e^{3\alpha}$

$\ln 3 = 3\alpha$

$\alpha = \frac{\ln 3}{3}$

$C(x) = 500\,e^{\frac{\ln 3}{3}x}$

$C(x) = 500\,(3)^{x/3}$

1-52

Find the area under the curve $y = 1/x$ from $x=1$ to $x=e^2$.

****

($e \doteq 2.7183$ is the base of the natural logarithms.)

The graph of the Region is shown below:

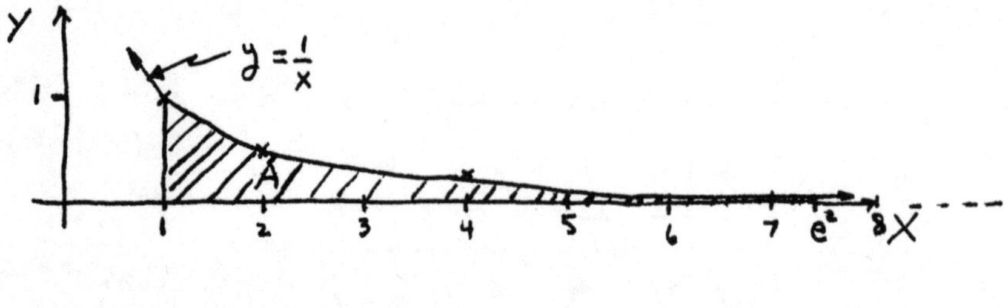

By definition $A = \int_1^{e^2} y\,dx = \int_1^{e^2} \left(\frac{1}{x}\right)dx = \operatorname{Ln} x \Big|_1^{e^2} = \operatorname{Ln} e^2 - \operatorname{Ln} 1.$

But $\operatorname{Ln} e^2 = 2\operatorname{Ln} e = 2(1) = 2$ and $\operatorname{Ln} 1 = 0$ so $A = 2 - 0 = \underline{\underline{2}}$ sq. units

1-53

Differentiate and simplify:

1. $y = \ln(x^3 - 2x)$ 2. $f(x) = x^{\sin x}$

****

1. According to the chain rule, $\frac{d}{dx}(\ln u) = \frac{1}{u}\frac{du}{dx}$. Then

$$\frac{dy}{dx} = \frac{1}{x^3-2x}\frac{d}{dx}(x^3-2x) = \frac{3x^2-2}{x^3-2x}$$

2. Use the fact that the exponential and logarithm are inverse functions to write

$$x^{\sin x} = e^{\ln x^{\sin x}} = e^{\sin x \ln x}$$

Then $\frac{df}{dx} = \frac{d}{dx} e^{\sin x \ln x} = e^{\sin x \ln x}\frac{d}{dx}(\sin x \ln x)$

$$= e^{\sin x \ln x}\left[\sin x\left(\frac{1}{x}\right) + \ln x(\cos x)\right]$$

$$= e^{\sin x \ln x}\left(\frac{1}{x}\sin x + \ln x \cos x\right)$$

$$= x^{\sin x}\left(\frac{1}{x}\sin x + \ln x \cos x\right)$$

In the last line, we replaced $e^{\sin x \ln x}$ by $x^{\sin x}$. Notice that the ordinary power rule, $\frac{d}{dx}(u^n) = nu^{n-1}\frac{du}{dx}$, does not apply to this problem, because the power $\sin x$ is not a constant.

1-54

Prove that $\int_0^1 \frac{dx}{x+(e-1)} = 1 - \ln(e-1)$

****

$$\int_0^1 \frac{dx}{x+(e-1)} = \ln\left[x+(e-1)\right]_0^1 = \ln[1+(e-1)] - \ln[0+(e-1)]$$

$$= \ln e - \ln(e-1) = 1 - \ln(e-1)$$

HARMONIC MOTION

1-55

An object is in simple harmonic motion with period $\frac{4\pi}{3}$, amplitude 2, and at time t = 0, x = 1. Find the equation of motion and the time at which the object attains maximum height.

****

$$x(t) = C \sin(\beta t + K) \qquad \frac{4\pi}{3} = \frac{2\pi}{\beta}$$

$$x(t) = 2 \sin\left(\frac{3}{2}t + K\right) \qquad \beta = \frac{3}{2}$$

$$1 = 2 \sin K$$

$$K = \frac{\pi}{6} \text{ OR } \frac{5\pi}{6}$$

$$x(t) = 2 \sin\left(\frac{3}{2}t + \frac{\pi}{6}\right) \quad \text{OR} \quad x(t) = 2 \sin\left(\frac{3}{2}t + \frac{5\pi}{6}\right)$$

$$2 = 2 \sin\left(\frac{3}{2}t + \frac{\pi}{6}\right) \qquad 2 = 2 \sin\left(\frac{3}{2}t + \frac{5\pi}{6}\right)$$

$$1 = \sin\left(\frac{3}{2}t + \frac{\pi}{6}\right) \qquad 1 = \sin\left(\frac{3}{2}t + \frac{5\pi}{6}\right)$$

$$\frac{3}{2}t + \frac{\pi}{6} = \frac{\pi}{2} \qquad \frac{3}{2}t + \frac{5\pi}{6} = \frac{\pi}{2}$$

$$t = \frac{2\pi}{9} \qquad t = -\frac{2\pi}{9}$$

1-56

Solve the vibrating spring equation $\frac{d^2y}{dt^2} = -4y$, where y = 0 at t = 0 and y = 10 at $t = \frac{\pi}{6}$.

****

The form of the given equation suggests cosines and sines as solutions since, for those functions, the second derivatives are directly proportional to the negative of the functions.

Set $y = A\cos ax + B\sin bx$

$y = 0$ at $t = 0$: $\quad 0 = A(1) + B(0) \qquad A = 0$

Then, $y = B\sin bx$

$y = 10$ at $t = \frac{\pi}{6}$: $\quad 10 = B\sin b\frac{\pi}{6}$ (Two unknowns)

$y = B\sin bx \qquad \frac{dy}{dx} = Bb\cos bx \qquad \frac{d^2y}{dx^2} = -Bb^2\sin bx$

$\frac{d^2y}{dt^2} = -4y$: $\quad -Bb^2\sin bx = -4B\sin bx$

$b^2 = 4 \qquad b = 2$

$$10 = B\sin\frac{b\pi}{6} = B\sin\frac{2\pi}{6} = B\sin\frac{\pi}{3} = B\frac{\sqrt{3}}{2}$$

$$B = \frac{20}{\sqrt{3}} = \frac{20\sqrt{3}}{3}$$

Thus, $y = \frac{20\sqrt{3}}{3}\sin 2x$ is the solution.

GROWTH AND DECAY

1-57

In an experiment, a tissue culture has been subjected to ionizing radiation. It was found that the number A of undamaged cells depends on the exposure time, in hours, according to the following formula

$$A = A_0 e^{kt}, \quad t \geq 0$$

If 5000 cells were present initially and 3000 survived a 2 hour exposure, find the elapsed time of exposure after which only half the original cells survive.

****

Since 5000 cells were present initially,

$$A = 5000e^{kt}.$$

So

$$3000 = 5000e^{k(2)}$$

$$3/5 = e^{2k}$$

$$\ln(3/5) = \ln(e^{2k}) = 2k\ln(e)$$

$$k = 1/2 \ln(3/5)$$

Thus,

$$A = 5000e^{1/2\ln(3/5)t}$$

Since 2500 is 1/2 the original number of cells,

$$2500 = 5000e^{1/2\ln(3/5)t}$$

$$1/2 = e^{1/2\ln(3/5)t}$$

$$\ln(1/2) = \ln(e^{1/2\ln(3/5)t}) = 1/2\ln(3/5)t \cdot \ln(e)$$

$$t = \frac{\ln(1/2)}{1/2\ln(3/5)} \approx 2.71 \text{ hours}$$

1-58

A lettuce leaf collected from the salad bar at the college cafeteria contains 99/100 as much carbon C^{14} as a freshly cut lettuce leaf. How old is it? (Use 5700 years for the half-life of C^{14}.)

****

LET $P(t)$ BE THE AMOUNT OF C^{14} AT TIME t. WE ASSUME THAT THE RATE OF CHANGE OF C^{14} IS PROPORTIONAL TO THE AMOUNT PRESENT. THAT IS,

$$\frac{dP}{dt} = kP.$$

SEPARATING VARIABLES AND INTEGRATING YIELDS

$$P(t) = P_0 e^{kt}$$

WHERE P_0 IS THE AMOUNT OF C^{14} AT $t=0$.

WE KNOW THAT $P(5700) = \frac{1}{2} P_0$

$$P_0 e^{5700k} = \frac{1}{2} P_0$$

$$\ln e^{5700k} = \ln \frac{1}{2} = -\ln 2$$

$$5700k = -\ln 2$$

$$k = \frac{-\ln 2}{5700}$$

SO $P(t) = P_0 e^{-\frac{\ln 2}{5700}t} = P_0 \left(e^{\ln 2}\right)^{\frac{-t}{5700}} = P_0\, 2^{\frac{-t}{5700}}$.

WE NEED TO KNOW WHEN $P(t) = \frac{99}{100} P_0$.

$$P_0\, 2^{\frac{-t}{5700}} = \frac{99}{100} P_0$$

$$\ln 2^{\frac{-t}{5700}} = \ln \frac{99}{100}$$

$$-\frac{t}{5700} \ln 2 = \ln 0.99$$

$$t = -\frac{\ln 0.99}{\ln 2}(5700) \doteq 82.65$$

THE LETTUCE AT THE CAFETERIA IS ABOUT 83 YEARS OLD.

1-59

Assume that the rate of growth of a population of fruit flies is proportional to the size of the population at each instant of time. If 100 fruit flies are present initially and 200 are present after 5 days, how many will be present after 10 days?

****

$N(t)$ represents the number of fruit flies present at a given time t.

GIVEN: $\frac{dN}{dt} = KN$, K is the growth constant

(1) $N(t) = N_0 e^{Kt}$ satisfies the given differential equation.

$N(10)$ is the information we are after.
$N(5) = 200$, GIVEN.
$N(0) \to N_0 = 100$, GIVEN.

USING (1) : $200 = 100e^{K5}$, $t=5$

$$2 = e^{5K},$$
$$\ln 2 = 5K$$
$$K = \frac{\ln 2}{5}$$

Substituting in (1)

$$N(t) = 100e^{\frac{\ln 2}{5}t}$$

$$\therefore \quad N(10) = 100e^{\frac{\ln 2}{5}(10)}$$
$$= 100e^{2\ln 2}$$
$$= 100e^{\ln 2^2}$$
$$= 100 \cdot 2^2$$
$$= 400 \text{ fruit flies}$$

1-60

A population of bacteria is known to grow exponentially. If 4 million are observed initially and 9 million after 2 days, how many will be present after 3 days?

****

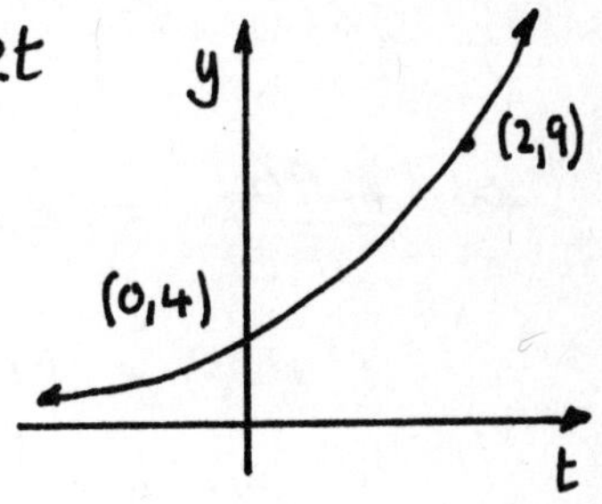

Assuming exponential growth $y = ce^{kt}$
We need to find the constants c and k.

(i) we know that $y=4$ when $t=0$.
that is

$$4 = ce^{k(0)} = ce^0 = c$$

(c is always the initial population)

Now we know $y = 4e^{kt}$

(ii) To find k use (2,9)

$$9 = 4e^{k(2)} \rightarrow \frac{9}{4} = e^{2k} \quad \text{take logs}$$

$$\ln\frac{9}{4} = \ln e^{2k} = 2k\ln e = 2k$$

thus $k = \frac{1}{2}\ln\frac{9}{4}$

So the growth equation becomes $y = 4e^{\frac{1}{2}\ln\frac{9}{4}.t}$

Thus to find the population at $t=3$

$$y = 4e^{\frac{1}{2}\ln\frac{9}{4}(3)} = 4e^{\frac{3}{2}\ln\frac{9}{4}} = 4e^{\ln\left(\frac{9}{4}\right)^{\left(\frac{3}{2}\right)}} \quad \text{using } \ln A^r = r\ln A$$

$$= 4e^{\ln\frac{27}{8}}$$

$$= 4 \cdot \frac{27}{8} \quad \text{using } e^{\ln x} = x$$

Thus the population of bacteria after 3 days will be $13\frac{1}{2}$ million

1-61

It takes money 20 years to triple at a certain rate of interest. How long does it take for money to double at this rate?

****

If $A(t)$ = amount of money at time t

$$\frac{dA(t)}{dt} = kA(t)$$ (because the rate of change is proportional to the amount)

so $$\frac{dA(t)}{A(t)} = k\,dt$$ We integrate both sides to get...

$$\ln A(t) = kt + c^*$$

$$A(t) = e^{kt + c^*}$$ If we let $c = e^{c^*}$

$$A(t) = ce^{kt}$$ Let $t = 0$. Then $A(0) = ce^0$, so $c = A(0)$. So...

$$A(t) = A(0)e^{kt}$$

To find k, we know that when $t = 20$, $A(t) = 3A(0)$.

So... $$3A(0) = A(0)e^{20k}$$

$$3 = e^{20k}$$

$$\ln 3 = 20k$$

so $$k = \frac{\ln 3}{20}$$

Thus $$A(t) = A(0)e^{\frac{\ln 3}{20}t}$$

Now, we want to find t when $A(t) = 2A(0)$.

$$2A(0) = A(0)e^{\frac{\ln 3}{20}t}$$

$$2 = e^{\frac{\ln 3}{20}t}$$

$$\ln 2 = \frac{\ln 3}{20}t$$

so $$t = \frac{20 \ln 2}{\ln 3} \text{ years}$$

Using a calculator, this is 12.62 years.

1-62

What annual rate of interest will make an investment of P dollars double in five years if the interest is compounded continuously?

****

The formula when interest is compounded continuously is

$A = Pe^{Rt}$, where e is the base of the natural logs. and \$A is the compounded amount after t years.

We are given the time for the investment to double. Hence in $t = 5$ years, $A = \$2P$. By substitution this yields

$2P = Pe^{5R}$. Now solve for R as follows:

Divide by P: $2 = e^{5R}$

Take LN of both sides: $\ln 2 = \ln e^{5R} = 5R$ ($\ln e = 1$)

Divide by 5:

Hence $R = \frac{\ln 2}{5}$ or $R \doteq 0.139$ ANS. $R \doteq 13.9\%$

1-63

Suppose that the number of bacteria in a culture at time t is given by $x = 5^4 e^{3t}$. Use natural logarithms to solve for t in terms of x.

****

Since $x = (5^4)(e^{3t})$, take the natural logarithm of both sides to get

$$\ln x = \ln[5^4(e^{3t})]$$

$$= 4\ln 5 + 3t \ln e$$

So $\ln x - 4\ln 5 = 3t$

So $\frac{\ln x - 4\ln 5}{3} = t$

1-64

In 1970, the Brown County groundhog population was 100. By 1980, there were 900 groundhogs in Brown County. If the rate of population growth of these animals is proportional to the population size, how many groundhogs might one expect to see in 1995?

****

Let $y = y(t)$ = population at time t (years)

$dy/dt = ky \Rightarrow y = Ce^{kt}$ for some constants C, k.

$y(0) = 100 = Ce^0 \Rightarrow C = 100$

$y(10) = 900 = 100e^{10k} \Rightarrow e^{10k} = 9$

$\therefore\ y(t) = 100(e^{10k})^{t/10} = 100(9)^{t/10}$

$y(25) = 100(9)^{25/10} = 100(9^{5/2}) = 100(243)$

$= 24{,}300$

$\therefore$ One might expect to see 24,300 groundhogs in Brown County by 1995.

1-65

Given $f'(t) = k \cdot f(t)$, $f(0) = 35$ and $f(3) = 945$, find $f(4)$.

****

$\frac{f'(t)}{f(t)} = K \Rightarrow \int \frac{f'(t)}{f(t)}\,dt = \int K\,dt \Rightarrow \ln f(t) = Kt + A \Rightarrow$

$f(t) = e^{Kt+A} = e^A \cdot e^{Kt}$. So $f(t) = Ce^{Kt}$ (where $C = e^a$).

$$f(0) = 35 \Rightarrow C = 35 \Rightarrow f(t) = 35e^{kt}.$$

$$f(3) = 945 \Rightarrow 35e^{3k} = 945 \Rightarrow e^{3k} = \frac{945}{35} = 27.$$

Therefore, $f(4) = 35e^{4k} = 35(e^{3k})^{\frac{4}{3}} = 35 \cdot 27^{\frac{4}{3}} = 35 \cdot 81 = 2835.$

1-66

In a certain medical treatment a tracer dye is injected into a human organ to measure its function rate and the rate of change of the amount of dye is proportional to the amount present at any time. If a physician injects 0.5g of dye and 30 minutes later 0.1 g remains, how much die will be present in 1½ hours?

****

The amount of dye present at any time, A, is

$$A = A_0 e^{kt}, \text{ where } A_0 \text{ is the initial}$$

amount of dye injected. So

$$A = 0.5e^{kt}, \text{ where } t \text{ is time in minutes.}$$

Also,

$$0.1 = 0.5e^{30k}$$
$$\tfrac{1}{5} = e^{30k}$$
$$\ln(\tfrac{1}{5}) = \ln(e^{30k}) = 30k \, \ln(e)^{\nearrow 1}$$
$$\therefore \quad k = \tfrac{1}{30}\ln(\tfrac{1}{5})$$

Thus, $A = 0.5e^{\tfrac{1}{30}\ln(\tfrac{1}{5})t}$

In 1½ hrs (90 minutes),

$$A = 0.5e^{90 \cdot \tfrac{1}{30}\ln(\tfrac{1}{5})} = 0.5e^{3\ln(\tfrac{1}{5})}$$
$$A \approx .004g.$$

1-67

John deposits $100 in a bank and at the same time Mary deposits $200. If John's bank pays 10% interest compounded continuously and Mary's bank pays 8% interest, how long must John wait till his bank account exceeds Mary's? [Use $\ln 2 \approx 0.7$]

****

Let $x(t)$ = John's money at time t
Let $y(t)$ = Mary's money at time t

then: $x(t) = 100 \exp(0.1t)$ & $y(t) = 200 \exp(0.08t)$

John's account will equal Mary's when:

$$100 \exp(0.1t) = 200 \exp(0.08t)$$

or: $\exp(0.1t) = 2 \exp(0.08t)$

taking natural logarithms on both sides:

$$0.1t = \ln 2 + 0.08t$$

so: $0.02t = \ln 2$

Hence $t = \dfrac{\ln 2}{.02} \approx \dfrac{.7}{.02} = \dfrac{70}{2} = 35$ years

For $t > 35$, $x(t) > y(t)$.

2 INTEGRATION TECHNIQUES

INTEGRATION BY PARTS

2-1

Evaluate the following integral $\int x\cot^{-1}(x)\,dx$

****

Using integration by parts, let $u = \cot^{-1}(x)$ and $dv = x\,dx$. Then $du = -\frac{1}{1+x^2}\,dx$ and $v = \frac{1}{2}x^2$

So $\int x\cot^{-1}(x)\,dx = \frac{1}{2}x^2\cot^{-1}(x) - \int -\frac{x^2}{2(1+x^2)}\,dx$

$$= \frac{1}{2}x^2\cot^{-1}(x) + \frac{1}{2}\int \frac{x^2}{1+x^2}\,dx$$

Since $\frac{x^2}{1+x^2} = 1 - \frac{1}{1+x^2}$,

$$\int x\cot^{-1}(x)\,dx = \frac{1}{2}x^2\cot^{-1}(x) + \frac{1}{2}\int \left[1 - \frac{1}{1+x^2}\right]dx$$

$$= \frac{1}{2}x^2\cot^{-1}(x) + \frac{1}{2}\int dx - \frac{1}{2}\int \frac{dx}{1+x^2}$$

$$= \frac{1}{2}x^2\cot^{-1}(x) + \frac{1}{2}x - \frac{1}{2}\tan^{-1}(x) + C$$

2-2

Evaluate the follwing integral.

$$\int \ln((e^x \ln x)^{1/x})\ dx$$

****

$$\int \ln(e^x \ln x)^{1/x}\, dx$$

$$= \int \frac{1}{x}\left[\ln(e^x \ln x)\right] dx$$

$$= \int \frac{1}{x} \ln e^x\, dx + \int \frac{1}{x} \ln(\ln x)\, dx$$

$$= \int \frac{1}{x} \cdot x\, dx + \int \frac{1}{x} \ln(\ln x)\, dx$$

$$= x + \int \frac{1}{x} \ln(\ln x) dx$$

substitute: $w = \ln x$
$dw = 1/x\, dx$

$$\int 1/x \ln(\ln x) dx$$

$$= \int \ln w\, dw$$

integrate by parts
$u = \ln w$
$dv = dw$
$du = \frac{1}{w} dw$
$v = w$

$$= w \ln w - \int dw$$

$$= w \ln w - w$$

$$= (\ln x)(\ln[\ln x]) - \ln x$$

$$= x + (\ln x)(\ln[\ln x]) - \ln x + C$$

2-3

Determine a reduction formula for $\int_1^e x(\ln x)^n dx$.

****

Call $I_n = \int_1^e x(\log x)^n dx$

using integration by parts $\int u\,dv = uv - \int v\,du$.
let $u = (\log x)^n \rightarrow du = n(\log x)^{n-1} \cdot \frac{1}{x}$
let $dv = x \rightarrow v = \frac{x^2}{2}$

substituting

$$I_n = \frac{x^2}{2}(\log x)^n \Big|_1^e - \int_1^e \frac{x^2}{2} \cdot n(\log x)^{n-1} \frac{1}{x}\,dx$$

$$= \frac{x^2}{2}(\log x)^n \Big|_1^e - \frac{n}{2}\int_1^e x(\log x)^{n-1}\,dx$$

$$= \frac{x^2}{2}(\log x)^n \Big|_1^e - \frac{n}{2} I_{n-1}$$

note $\log 1 = 0$.

and I_{n-1} is used for $\int_1^e x(\log x)^{n-1}\,dx$

$$\text{so } I_n = \frac{e^2}{2}(\log e)^n - \frac{n}{2} I_{n-1}$$

$$\underline{I_n = \frac{e^2}{2} - \frac{n}{2} I_{n-1}}$$

2-4

Find $\int x^2 e^{2x}\, dx$

****

Using the method of integration by parts, we let

$u = x^2$	$dv = e^{2x}\, dx$
or $\frac{du}{dx} = 2x$	$\frac{dv}{dx} = e^{2x}$
$du = 2x\, dx$	$v = \frac{1}{2} e^{2x}$

Substituting these in the given function, we have

$$\int x^2 e^{2x}\, dx = x^2\left(\frac{1}{2} e^{2x}\right) - \int\left(\frac{1}{2} e^{2x}\right) 2x\, dx$$

$$= \frac{1}{2} x^2 e^{2x} - \int x e^{2x}\, dx$$

For the integral on the right side we again integrate by parts and let

$u = x$	$dv = e^{2x}\, dx$
or $\frac{du}{dx} = 1$	$\frac{dv}{dx} = e^{2x}$
$du = dx$	$v = \frac{1}{2} e^{2x}$

Substituting these for the integral on the right side, we have

$$\int x e^{2x}\, dx = x\left(\frac{1}{2} e^{2x}\right) - \int\left(\frac{1}{2} e^{2x}\right) dx$$

The integral on the right side now yields

$$\int \frac{1}{2} e^{2x}\, dx = \frac{1}{4} e^{2x} + C$$

Hence, we write:

$$\int x^2 e^{2x}\, dx = \frac{1}{2} x^2 e^{2x} - \left[x\left(\frac{1}{2} e^{2x}\right) - \left(\frac{1}{4} e^{2x} + C\right)\right]$$

$$= \frac{1}{2} x^2 e^{2x} - \frac{1}{2} x e^{2x} + \frac{1}{4} e^{2x} + C$$

Ans.

2-5

Evaluate the following indefinite integral $\int [\ln(x)]^3 dx$

****

Using integration by parts, let $u = [\ln(x)]^3$ and $dv = dx$. Then $du = 3[\ln(x)]^2 \cdot \frac{1}{x} dx$ and $v = x$.

$$\int [\ln(x)]^3 = x[\ln(x)]^3 - \int 3[\ln(x)]^2 \cdot \frac{1}{x} \cdot x\, dx$$
$$= x[\ln(x)]^3 - 3\int [\ln(x)]^2 dx$$

Now let $\bar{u} = [\ln(x)]^2$ and $d\bar{v} = dx$. Then $d\bar{u} = 2\ln(x) \cdot \frac{1}{x} \cdot dx$ and $v = x$

$$\int [\ln(x)]^3 = x[\ln(x)]^3 - 3\left[x[\ln(x)]^2 - \int 2\ln(x) \cdot \frac{1}{x} \cdot x\, dx\right]$$
$$= x[\ln(x)]^3 - 3x[\ln(x)]^2 + 6\int \ln(x)\, dx$$

Now let $\bar{\bar{u}} = \ln(x)$ and $d\bar{\bar{v}} = dx$. Then $d\bar{\bar{u}} = \frac{1}{x} dx$ and $\bar{\bar{v}} = x$

$$\int [\ln(x)]^3 = x[\ln(x)]^3 - 3x[\ln(x)]^2 + 6\left[x\ln(x) - \int \frac{1}{x} \cdot x\, dx\right]$$
$$= x[\ln(x)]^3 - 3x[\ln(x)]^2 + 6x\ln(x) - 6\int dx$$
$$= x[\ln(x)]^3 - 3x[\ln(x)]^2 + 6x\ln(x) - 6x + C$$

2-6

Evaluate: $\int_1^e (\ln x)^2\, dx$

****

integrate by parts:

$u = (\ln x)^2 \qquad dv = dx$

$du = \frac{2(\ln x)}{x}\, dx \qquad v = x$

$$\int_1^e (\ln x)^2 dx = x(\ln x)^2 - \int_1^e (x)\frac{2\ln x}{x}\, dx$$

$$= x(\ln x)^2 - 2\int_1^e \ln x\, dx$$

$u = \ln x \qquad dv = dx$

$du = \frac{dx}{x} \qquad v = x$

$$= x(\ln x)^2 - 2\left(x\ln x - \int_1^e \frac{x\, dx}{x}\right)$$

$$= x(\ln x)^2 - 2x\ln x + 2\int_1^e dx$$

$$= \left(x(\ln x)^2 - 2x\ln x + 2x\right)_1^e$$

$$= \left(e(\ln e)^2 - 2e\ln e + 2e\right) - \left(1(\ln 1)^2 - 2(1)\ln 1 + 2\right)$$

$$= e - 2e + 2e - 2 = e - 2$$

2-7

Find the area of the region bounded by the curve $y = \tan^{-1} x$, the x-axis and the line $x = 1$.

****

Sketch of the region:

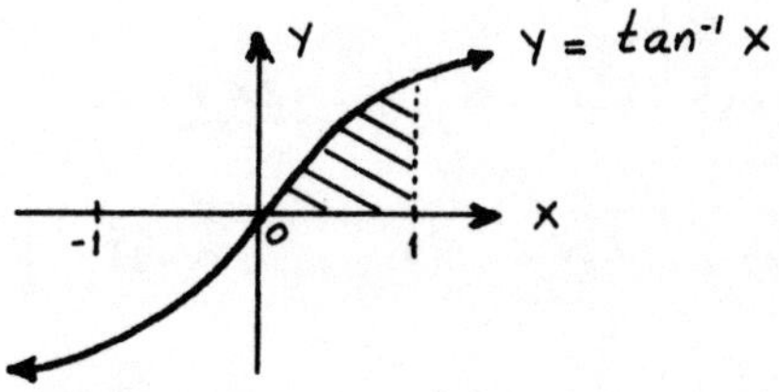

It is necessary to evaluate: $\int_0^1 \tan^{-1} x \, dx$

This will be done by parts.

Let $u = \tan^{-1} x$ and $dv = dx$.

Then $\frac{du}{dx} = \frac{1}{1+x^2} \Rightarrow du = \frac{dx}{1+x^2}$ and $v = x$.

So $$\int_0^1 \tan^{-1} x \, dx = x \tan^{-1} x \Big]_0^1 - \int_0^1 \frac{x\,dx}{1+x^2}$$

$$= \tan^{-1} 1 - \left(\tfrac{1}{2} \ln |1+x^2|\right)\Big]_0^1$$

$$= \frac{\pi}{4} - \frac{\ln 2}{2} = \frac{\pi - \ln 4}{4}$$

2-8 ■■■

Integrate: $\int \dfrac{LN2xdx}{x}$ (x>o)

****

Note that the integrand $\dfrac{LN\,2X}{X} = \dfrac{LN2 + LNX}{X} = \dfrac{LN2}{X} + \dfrac{LNX}{X}$

By substitution, $\int \dfrac{LN2X\,dx}{X} = \int \dfrac{LN2}{X}\,dx + \int \dfrac{LNX}{X}\,dx$.

Since $LN2$ is a constant, $\int \dfrac{LN2}{X}\,dx = LN2 \int \dfrac{dx}{X} = LN2(LNX)$

FoR $\int \dfrac{LNX}{X}\,dx$ use the parts formula $\int u\,dv = uv - \int v\,du$.

Choose: $u = LNX$, $dv = \dfrac{dx}{X}$

Then $du = \dfrac{dx}{X}$, $v = LNX$ and

$\int \dfrac{LNX\,dx}{X} = LN^2X - \int \dfrac{LNX}{X}\,dx$. Solving algebraically

yields $2\int \dfrac{LNX\,dx}{X} = LN^2X$ oR $\int \dfrac{LNX}{X}\,dx = \dfrac{1}{2}LN^2X$

Finally by back substitution we have

$\int \dfrac{LN2X}{X}\,dx = LN2\,(LNX) + \dfrac{1}{2}LN^2X + C$ $\underline{\underline{ANS.}}$

2-9 ■■■

Find $\int x^2 \cos x\,dx$.

****

Using integration by parts (twice),

$\int x^2 \cos x\,dx = x^2 \sin x - \int 2x \sin x\,dx = x^2 \sin x - \left[-2x\cos x + 2\int \cos x\,dx\right]$

$u = x^2 \quad dV = \cos x\,dx \qquad u = 2x \quad dV = \sin x\,dx$

$du = 2x\,dx \quad V = \sin x \qquad du = 2dx \quad V = -\cos x$

$= x^2 \sin x + 2x\cos x - 2\sin x + C.$

2-10

Integrate $\int \sec^3 x \, dx$

****

$$\int \sec^3 x \, dx = \int \sec^2 x \sec x \, dx$$

$$u = \sec x \qquad dv = \sec^2 x \, dx$$
$$du = \sec x \tan x \, dx \qquad v = \tan x$$

Using integration by parts:

$$I = \sec x \tan x - \int \tan^2 x \sec x \, dx$$
$$= \sec x \tan x - \int (\sec^2 x - 1) \sec x \, dx$$
$$= \sec x \tan x - \int \sec^3 x \, dx + \int \sec x \, dx$$
$$= \sec x \tan x - I + \int \sec x \, dx$$

$$\Rightarrow 2I = \sec x \tan x + \int \sec x \, dx$$

Recalling that $\int \sec x \, dx = \ln|\sec x + \tan x|$

$$\Rightarrow 2I = \sec x \tan x + \ln|\sec x + \tan x| + C$$

$$\Rightarrow I = \frac{\sec x \tan x}{2} + \frac{\ln|\sec x + \tan x|}{2} + C'$$

where $C' = \frac{C}{2}$

2-11

Find $\int x^3 e^{3x} dx$.

****

We will apply Integration by Parts three times in order to complete this problem.

$$\int x^3 e^{3x} dx = \frac{1}{3}x^3 e^{3x} - \frac{3}{3}\int x^2 e^{3x} dx =$$

Let $v = x^3 \Rightarrow dv = 3x^2 dx$ Let $v = x^2 \Rightarrow dv = 2x\, dx$

$du = e^{3x} dx \Rightarrow u = \frac{1}{3}e^{3x}$ $du = e^{3x} dx \Rightarrow u = \frac{1}{3}e^{3x}$

$$\frac{1}{3}x^3 e^{3x} - \left\{ \frac{1}{3}x^2 e^{3x} - \frac{2}{3}\int x e^{3x} dx \right\} =$$

Let $v = x \Rightarrow dv = dx$

$du = e^{3x} dx \Rightarrow u = \frac{1}{3}e^{3x}$

$$\frac{1}{3}x^3 e^{3x} - \frac{1}{3}x^2 e^{3x} + \frac{2}{9}x e^{3x} - \frac{2}{9}\int e^{3x} dx =$$

$$\frac{1}{3}x^3 e^{3x} - \frac{1}{3}x^2 e^{3x} + \frac{2}{9}x e^{3x} - \frac{2}{27}e^{3x} + C.$$

2-12

Find $\int xe^{3x}\, dx$.

****

Use Parts: $\int u\, dv = uv - \int v\, du$

Let $u = x$ and $dv = e^{3x} dx$

$du = dx$ $v = \frac{1}{3}e^{3x}$

So $\int x e^{3x} dx = \frac{1}{3}x e^{3x} - \int \frac{1}{3} e^{3x} dx$

$$= \frac{1}{3}x e^{3x} - \frac{1}{9}e^{3x} + C$$

2-13

Find $\int x^3 e^{x^2}\, dx$

****

Note that $\int x e^{x^2} dx = \frac{1}{2} e^{x^2} + c$ (let $w = x^2$).

Integrate by parts: $u = x^2$ $\quad dv = x e^{x^2} dx$

$du = 2x\, dx$ $\quad v = \frac{1}{2} e^{x^2} + c$

$$\int x^3 e^{x^2} dx = \frac{1}{2} x^2 e^{x^2} - \int x e^{x^2} dx$$

$$= \frac{1}{2} x^2 e^{x^2} - \frac{1}{2} e^{x^2} + c.$$

2-14

Evaluate the following indefinite integral:

$$\int x^2 \cos 2x\, dx$$

****

$$\int x^2 \cos 2x\, dx = \frac{x^2}{2} \sin 2x - \int x \sin 2x\, dx$$

let $u = x^2$, $du = 2x\, dx$ $\quad$ let $s = x$, $ds = dx$

$\cos 2x\, dx = dv$, $v = \frac{1}{2} \sin 2x$ $\quad$ $\sin 2x\, dx = dt$, $t = -\frac{1}{2} \cos 2x$

$$= \frac{x^2}{2} \sin 2x - \left[-\frac{x}{2} \cos 2x - \int -\frac{1}{2} \cos 2x\, dx \right]$$

$$= \frac{x^2}{2} \sin 2x + \frac{x}{2} \cos 2x - \frac{1}{4} \sin 2x + C$$

INTEGRATION BY SUBSTITUTION

2-15

Evaluate the following indefinite integral:

$$\int \frac{4\sqrt{x}}{6 + x}\, dx$$

****

Let $z^2 = x$, then $z = \sqrt{x}$ and $2z\,dz = dx$.

Substituting $\int \frac{4\sqrt{x}}{6+x}\, dx = \int \frac{4z}{6+z^2} \cdot 2z\,dz$

$$= 8\int \frac{z^2\,dz}{6+z^2} = 8\int \left[1 - \frac{6}{6+z^2}\right] dz$$

$$= 8\int dz - 48\int \frac{dz}{6+z^2}$$

$$= 8z - 48 \cdot \frac{1}{\sqrt{6}} \tan^{-1}\left(\frac{z}{\sqrt{6}}\right) + C$$

$$= 8\sqrt{x} - 8\sqrt{6}\tan^{-1}\left(\frac{\sqrt{x}}{\sqrt{6}}\right) + C$$

$$= 8\sqrt{x} - 8\sqrt{6}\tan^{-1}\left(\sqrt{\frac{x}{6}}\right) + C$$

2-16

Evaluate the following integrals:

(a) $\int_0^{1/4} \sec(\pi u)\ \tan(\pi u)\ du$

(b) $\int x\ (1 + x^3)^2\ dx$

****

(a) By substitution: $\omega = \pi u \quad d\omega = \pi\, du$

$u = 0 \Rightarrow \omega = 0, \quad u = 1/4 \Rightarrow \omega = \pi/4$

$$\text{integral} = \frac{1}{\pi}\int_0^{\pi/4} \sec\omega\ \tan\omega\ d\omega$$

$$= \frac{1}{\pi}\sec\omega\Big|_0^{\pi/4} = \frac{1}{\pi}(\sqrt{2} - 1)$$

(b) $$\text{integral} = \int x\,(1 + 2x^3 + x^6)\,dx$$

$$= \int (x + 2x^4 + x^7)\,dx = \frac{x^2}{2} + \frac{2x^5}{5} + \frac{x^8}{8} + C$$

2-17

$\int \sin^5 x \cos x\ dx$ is (a) $1/12 \sin^6 x \cos^2 x + C$ (b) $1/6 \sin^6 x \cos x + C$ (c) $-\sin^6 x + 5\sin^4 x \cos^2 x + C$ (d) $1/6 \sin^6 x + C$ (e) $1/6 \sin^7 x + C$.

****

$$\int \sin^5 x \cos x\,dx = \int u^5\,du = \frac{1}{6}u^6 + C = \frac{1}{6}\sin^6 x + C.$$

let $u = \sin x$, $du = \cos x\,dx$

2-18

Integrate:

1. $\int \frac{dx}{\sqrt{9 - 4x^2}}$

2. $\int t \sin t \, dt$

****

1. This can be transformed to the form $\int \frac{du}{\sqrt{1-u^2}} = \operatorname{Sin}^{-1} u + C$. First, factor 9 from the radical.

$$\int \frac{dx}{\sqrt{9-4x^2}} = \int \frac{dx}{\sqrt{9(1 - \frac{4}{9}x^2)}} = \int \frac{dx}{3\sqrt{1-\frac{4}{9}x^2}}$$

Now multiply and divide by $\frac{2}{3}$ and let $u = \frac{2}{3}x$, $du = \frac{2}{3}dx$.

$$\int \frac{dx}{3\sqrt{1 - \frac{4}{9}x^2}} = \int \frac{\frac{2}{3}dx}{(\frac{2}{3})(3)\sqrt{1-\frac{4}{9}x^2}} = \int \frac{\frac{2}{3}dx}{2\sqrt{1-(\frac{2}{3}x)^2}}$$

$$= \frac{1}{2}\int \frac{\frac{2}{3}\,dx}{\sqrt{1-(\frac{2}{3}x)^2}} = \frac{1}{2}\operatorname{Sin}^{-1}(\tfrac{2}{3}x) + C$$

2. Integration by parts is needed. Let $u = t$, $dv = \sin t$. Then $du = dt$, $v = -\cos t$, and

$$\int t \sin t \, dt = -t\cos t + \int \cos t \, dt = -t\cos t + \sin t$$

2-19

Find $\int_0^3 \frac{x^3\, dx}{\sqrt{16 + x^2}}$ by making a nontrigonometric substitution.

****

Let $u = \sqrt{16 + x^2} = (16 + x^2)^{1/2}$

Hence, we have $u^2 = 16 + x^2$ or $x^2 = u^2 - 16$

Also, $\frac{du}{dx} = \frac{1}{2}(16 + x^2)^{-1/2}\, 2x$

$du = \frac{x \cdot dx}{\sqrt{16 + x^2}}$

$= \frac{x \cdot dx}{u}$

or, $u \cdot du = x \cdot dx$

when $x = 3$

$u = \sqrt{16 + 3^2} = 5$

when $x = 0$

$u = \sqrt{16 + 0^2} = 4$

By substitution, therefore, we have

$$\int_0^3 \frac{x^3\, dx}{\sqrt{16 + x^2}} = \int_0^3 \frac{x^2\, x\, dx}{\sqrt{16 + x^2}}$$

$$= \int_4^5 \frac{(u^2 - 16)\, u\, du}{u}$$

$$= \int_4^5 (u^2 - 16)\, du$$

$$= \frac{u^3}{3} - 16u \Big/_4^5$$

$$= \left[\frac{(5)^3}{3} - 16(5)\right] - \left[\frac{(4)^3}{3} - 16(4)\right]$$

$$= \frac{125}{3} - 80 - \frac{64}{3} + 64$$

$$= \frac{13}{3} \quad \text{Ans.}$$

2-20

$$\int \frac{x^3+2x}{\sqrt{x^2-2}}\,dx =$$

****

$$\int \frac{x^3+2x}{\sqrt{x^2-2}} = \int \frac{x(x^2+2)}{\sqrt{x^2-2}}\,dx. = I$$

$u = x^2-2 \leftrightarrow x^2 = u+2$; so $x^2+2 = u+4$.

$du = 2x\,dx$

$$I = \int \frac{u+4}{\sqrt{u}}\,\frac{du}{2} = \frac{1}{2}\int \left(\sqrt{u} + \frac{4}{\sqrt{u}}\right)du$$

$$= \frac{1}{2}\int \left(u^{1/2} + 4u^{-1/2}\right)du = \frac{1}{2}\left(\frac{2}{3}u^{3/2} + 8u^{1/2}\right)$$

SUBSTITUTE

$$I = \frac{1}{3}(x^2-2)^{3/2} + 4(x^2-2)^{1/2} + C.$$

2-21

$$\int \frac{e^x+e^{-x}}{e^x-e^{-x}}\,dx =$$

****

$$\int \frac{e^x+e^{-x}}{e^x-e^{-x}}\,dx = \int \frac{du}{u} = \ln u + C.$$

$u = e^x - e^{-x}$

$du = (e^x + e^{-x})\,dx$

$$= \ln(e^x + e^{-x}) + C.$$

INTEGRATION OF RATIONAL FUNCTIONS

2-22

Find $\int \frac{x^2}{x^2 + 4x + 5} \, dx$

****

By long division,

$$\int \frac{x^2}{x^2+4x+5} \, dx = \int \left(1 - \frac{4x+5}{x^2+4x+5}\right) dx$$

$$= \int 1 \, dx - \int \frac{4x+5}{x^2+4x+5} \, dx$$

(1) $\int 1 \, dx = x$

(2) $\int \frac{4x+5}{x^2+4x+5} \, dx = \int \frac{4x+5}{(x+2)^2+1} \, dx$

(Now let $u = x+2$, so $x = u-2$)

$$= \int \frac{4u-3}{u^2+1} \, du = 4\int \frac{u \, du}{u^2+1} - 3\int \frac{du}{u^2+1}$$

$$= 2 \ln(u^2+1) - 3 \arctan u$$

$$= 2 \ln(x^2+4x+5) - 3 \arctan(x+2)$$

(3) Answer $= x - 2\ln(x^2+4x+5) + 3\arctan(x+2) + c$

2-23 ■■

Perform the following integration:

$$\int \frac{x^4 + 3x^2 + 8x + 9}{x^2 + 4} \, dx$$

****

using long division, simplify the integrand

$$
\begin{array}{r}
x^2 - 1 \\
x^2+4 \overline{\smash{\big)}\ x^4+3x^2+8x+9} \\
\underline{x^4+4x^2} \\
-x^2 + 8x + 9 \\
\underline{-x^2 \qquad -4} \\
8x + 13
\end{array}
$$

$$\int \frac{x^4 + 3x^2 + 8x + 9}{x^2 + 4} \, dx = \int x^2 - 1 + \frac{8x + 13}{x^2 + 4} \, dx$$

$$= \int x^2 - 1 + \underbrace{\frac{8x}{x^2 + 4}}_{\substack{\downarrow \\ u = x^2+4 \\ du = 2x \, dx \\ \text{form: } \int \frac{du}{u}}} + \underbrace{\frac{13}{x^2 + 4}}_{\substack{\downarrow \\ u = x \\ a = 2 \\ \text{form: } \int \frac{du}{u^2 + a^2}}} \, dx$$

$$= \frac{x^3}{3} - x + 4 \ln|x^2 + 4| + \frac{13}{2} \arctan(x/2) + C$$

2-24

Evaluate $\int \frac{2x + 1}{x^2 + 4x + 5}\, dx$

****

$$\int \frac{2X+1}{X^2+4X+5}\,dx = \int \frac{2X+4}{X^2+4X+5}\,dx + \int \frac{-3}{X^2+4X+5}\,dx$$

$$= \int \frac{2X+4}{X^2+4X+5}\,dx - 3\int \frac{1}{(X+2)^2+1}\,dx$$

$$= \ln|X^2+4X+5| - 3 \arctan(X+2) + C$$

2-25

Evaluate $\int \left(\frac{x}{(x^2 + 1)^2} + \frac{1}{x^2 + 1}\right) dx.$

****

$$\int \left(\frac{x}{(x^2+1)^2} + \frac{1}{x^2+1}\right) dx =$$

$$\int \frac{x}{(x^2+1)^2}\,dx + \int \frac{1}{x^2+1}\,dx$$

↑ Substitute $u = x^2+1$, $du = 2x\,dx$, $\frac{1}{2}du = x\,dx$

↑ get $\tan^{-1}x + c$

$$= \int \frac{1}{u^2}\cdot\frac{1}{2}\,du + \tan^{-1}x + c = -\frac{1}{2}u^{-1} + \tan^{-1}x + c$$

$$= -\frac{1}{2}(x^2+1)^{-1} + \tan^{-1}x + c.$$

NUMERICAL INTEGRATION

2-26

Evaluate $\int_0^2 \sqrt{x^3 + 2}\, dx$ using Simpson's Rule with n = 6.

****

$$A_S = \frac{\Delta X}{3}(Y_0 + 4Y_1 + 2Y_2 + 4Y_3 + 2Y_4 + 4Y_5 + Y_6)$$

$f(x) = \sqrt{x^3+2}$ $\qquad \Delta X = \frac{b-a}{n} = \frac{2-0}{6} = 1/3$

$Y_0 = f(0) = 1.414 \times 1 = 1.414$

$Y_1 = f(1/3) = 1.427 \times 4 = 5.708$

$Y_2 = f(2/3) = 1.515 \times 2 = 3.030$

$Y_3 = f(1) = 1.732 \times 4 = 6.928$

$Y_4 = f(4/3) = 2.091 \times 2 = 4.182$

$Y_5 = f(5/3) = 2.575 \times 4 = 10.300$

$Y_6 = f(2) = 3.162 \times 1 = \underline{3.162}$

34.724

$A_S = \frac{1/3}{3}(34.724)$

$\approx \underline{\underline{3.86}}$

2-27

Use Simpson's rule with n = 10 to approximate $\int_0^1 \frac{1}{1 + x^2}\,dx$

****

Simpsons rule $\frac{\Delta x}{3}\left[f x_0 + 2(f x_2 + \cdots) + 4(f x_1 + \cdots) + f x_n\right]$

where $\Delta x = \frac{b-a}{n} = \frac{1-0}{10} = .1$

Divide $\frac{1}{1+x^2}$ into 10 equal intervals between x=0 and x=1

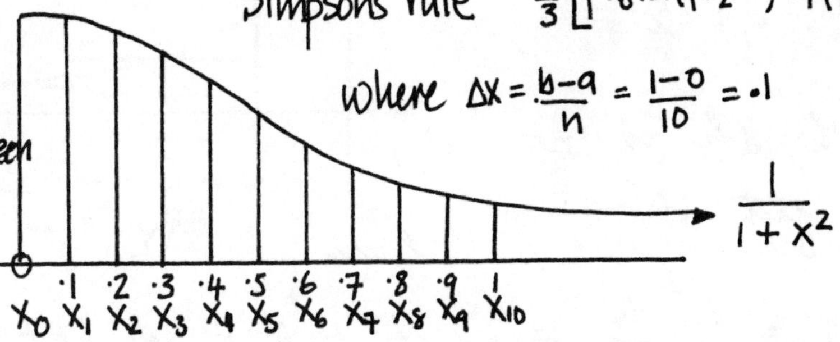

Identify the ordinates $x_0 - x_{10}$, and compute the function value for these ordinates

x_0	0	$fx_0 = 1$
x_1	·1	$fx_1 = .9901$
x_2	·2	$fx_2 = .9615$
x_3	·3	$fx_3 = .9174$
x_4	·4	$fx_4 = .8621$
x_5	·5	$fx_5 = .8$
x_6	·6	$fx_6 = .7353$
x_7	·7	$fx_7 = .6711$
x_8	·8	$fx_8 = .6096$
x_9	·9	$fx_9 = .5525$
x_{10}	1	$fx_{10} = .5$

$$\text{Area} \sim \frac{.1}{3}\left[f_0 + 2(f_2 + f_4 + f_6 + f_8) + 4(f_1 + f_3 + f_5 + f_7 + f_9) + f_{10}\right]$$

$$= \frac{.1}{3}\left[1 + 2(.9615 + .8621 + .7353 + .6096) + 4(.9901 + .9174 + .8 + .6711 + .5525) + .5\right]$$

$$= .0333\left[1 + 2(3.1685) + 4(3.9311) + .5\right]$$

$$= .0333\left[1 + 6.337 + 15.7244 + .5\right] = .0333\left[23.5614\right]$$

$$= 0.7846.$$

So $\int_0^1 \frac{1}{1+x^2}\,dx \sim 0.7846.$

It can be shown that $\int_0^1 \frac{1}{x^2+1}\,dx = \frac{\pi}{4}$ thus $(0.7846) \times 4$ can be used as an approximation to π. The more ordinates we take the more accurate the approximation

2-28 ■■■

Use the midpoint rule with 2 equal subdivisions to get an approximation for ln 5.

****

By definition,

$$\ln 5 = \int_1^5 \frac{dx}{x}$$

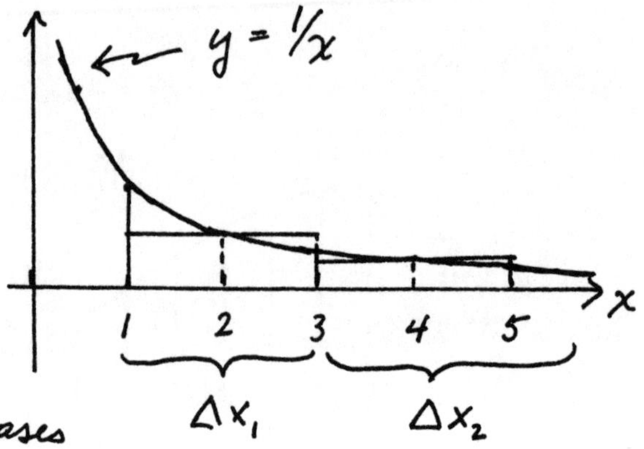

As the sketch to the right shows:

$$\ln 5 = \int_1^5 \frac{dx}{x} \approx 2 \cdot \frac{1}{2} + 2 \cdot \frac{1}{4} = 1 + \frac{1}{2} = 1.5$$

bases

heights of rectangles

$y = 1/x$

$\Delta x_1 \qquad \Delta x_2$

INTEGRALS INVOLVING
TRIGONOMETRIC FUNCTIONS

2-29 ■■■

Integrate $\displaystyle\int \frac{\tan^2(x)}{\cos^2(x)}\, dx$

****

$$\int \frac{\tan^2 x}{\cos^2 x}\, dx = \int (\tan x)^2 (\sec^2 x\, dx) \qquad \text{let } u = \tan x$$
$$du = \sec^2 x\, dx$$
$$= \int u^2\, du$$
$$= \frac{1}{3} u^3 + C = \frac{1}{3} \tan^3 x + C$$

2-30

Evaluate: $\int_{\pi/6}^{\pi/2} \cos^3 x \sqrt{\sin x}\ dx$

****

$$\int_{\pi/6}^{\pi/2} \cos^3 x (\sin x)^{1/2} dx = \int_{\pi/6}^{\pi/2} (\cos^2 x)(\sin x)^{1/2}(\cos x\, dx)$$

$$= \int_{\pi/6}^{\pi/2} (1 - \sin^2 x)(\sin x)^{1/2}(\cos x\, dx)$$

$$= \int_{\pi/6}^{\pi/2} (\sin x)^{1/2}(\cos x\, dx) - (\sin x)^{5/2}(\cos x\, dx)$$

$$= \frac{2}{3}(\sin x)^{3/2} - \frac{2}{7}(\sin x)^{7/2} \Big]_{\pi/6}^{\pi/2}$$

$$= \left[\frac{2}{3}(\sin \pi/2)^{3/2} - \frac{2}{7}(\sin \pi/2)^{7/2}\right] - \left[\frac{2}{3}(\sin \pi/6)^{3/2} - \frac{2}{7}(\sin \pi/6)^{7/2}\right]$$

$$= \frac{2}{3} - \frac{2}{7} - \frac{2}{3}\left(\frac{1}{2}\right)^{3/2} + \frac{2}{7}\left(\frac{1}{2}\right)^{7/2} = \frac{32\sqrt{2} - 25}{84\sqrt{2}}$$

2-31

Find $\int \sin^4 x \cos^3 x\ dx$.

****

$$\int \sin^4 x (1 - \sin^2 x)[\cos x\, dx] = \int u^4 (1 - u^2)\, du$$

let $u = \sin x$

$du = \cos x\, dx$

$$= \int [u^4 - u^6]\, du$$

$$= \frac{1}{5}u^5 - \frac{1}{7}u^7 + C$$

$$= \frac{1}{5}\sin^5 x - \frac{1}{7}\sin^7 x + C.$$

2-32

$$\int \frac{\sin^5 x + \sin^3 x}{\cos^4 x}\, dx =$$

****

$$\int \frac{\sin^5 x + \sin^3 x}{\cos^4 x} = \int \frac{\sin^3 x(\sin^2 x + 1)}{\cos^4 x}\, dx. = I$$

$$u = \cos x.$$

$$du = -\sin x\, dx.$$

Since $u = \cos x$ $\quad \sin^2 x = (1-u^2)$

so

$$I = \int \frac{(1-u^2)(2-u^2)}{u^4}\, du = \int \frac{2 - 3u^2 + u^4}{u^4}\, du$$

$$= \int (2u^{-4} - 3u^{-2} + 1)\, du = -\frac{2}{3}u^{-3} + 3u^{-1} + u$$

SUBSTITUTE BACK

$$I = -\frac{2}{3\cos^3 x} + \frac{3}{\cos x} + \cos x + C.$$

2-33

Find $\int \sec^4(3x)dx$

****

$$\int \sec^4(3x)dx = \int \sec^2(3x)(1+\tan^2(3x))\,dx$$

$=$ (If we let $u = \tan(3x)$ so

$du = 3\sec^2(3x)dx$)

$$\frac{1}{3}\int (1+u^2)du$$

$$= \frac{u}{3} + \frac{u^3}{9} + C$$

$$= \frac{\tan(3x)}{3} + \frac{\tan^3(3x)}{9} + C$$

2-34

Evaluate $\int \cos(2x)\tan(x)\ dx$

****

$$\int \cos 2x \tan x\, dx = \int (2\cos^2 x - 1)\frac{\sin x}{\cos x}\,dx$$

$$= \int (2\sin x\cos x - \tan x)\,dx$$

$$= \int (\sin 2x - \tan x)\,dx$$

$$= -\frac{1}{2}\cos 2x + \ln|\cos x| + C$$

2-35

Find the area under the curve $y = 5\sec^2(2x)$ and above the interval $[\frac{\pi}{8}, \frac{\pi}{6}]$.

****

$$\text{Area} = \int_{\pi/8}^{\pi/6} 5\sec^2(2x)\,dx \quad (\text{by sub. } u = 2x,\ du = 2dx)$$

$$= \frac{5}{2}\int_{\pi/4}^{\pi/3} \sec^2 u\,du = \frac{5}{2}\tan u\Big|_{\pi/4}^{\pi/3} = \frac{5}{2}(\sqrt{3} - 1)$$

2-36

Integrate: $\displaystyle\int \frac{dx}{\sin \frac{1}{2}x}$

****

$$\int \frac{dx}{\sin\frac{1}{2}x} = \int \csc\tfrac{1}{2}x\,dx$$

$$= 2\int \tfrac{1}{2}\csc\tfrac{1}{2}x\,dx$$

$$= 2\ln\left|\csc\tfrac{1}{2}x - \cot\tfrac{1}{2}x\right| + C$$

2-37

Integrate: $\int \cos^3 t\, dt$

****

Rewrite the integrand $\cos^3 t$ as follows: $\cos^3 t = \cos t\,(\cos^2 t) = \cos t\,(1 - \sin^2 t) = \cos t - \sin^2 t \cos t$. Then

$\int \cos^3 t\, dt = \int \cos t\, dt - \int \sin^2 t \cos t\, dt$, and $\int \sin^2 t \cos t\, dt = \int u^2 du$ where $u = \sin t$ and $du = \cos t\, dt$. Hence by the power Rule $\int \sin^2 t \cos t\, dt = \frac{u^3}{3} + c = \frac{\sin^3 t}{3} + c$ and

finally $\int \cos^3 t\, dt = \sin t - \frac{\sin^3 t}{3} + c$.

2-38

Evaluate the integral $\int \sqrt{\cos x}\, \sin^3 x\, dx$.

****

$$\int \sqrt{\cos x}\, \sin^3 x\, dx = \int (\cos x)^{\frac{1}{2}} \sin^2 x \sin x\, dx$$

$$= \int (\cos x)^{\frac{1}{2}} (1 - \cos^2 x) \sin x\, dx$$

$$= \int \left[(\cos x)^{\frac{1}{2}} - (\cos x)^{\frac{5}{2}}\right] \sin x\, dx$$

$$= -\frac{2}{3}(\cos x)^{\frac{3}{2}} + \frac{2}{7}(\cos x)^{\frac{7}{2}} + C$$

INTEGRATION BY TRIGONOMETRIC SUBSTITUTION

2-39

Perform the following integration:

$$\int \frac{3x^2\,dx}{\sqrt{4 - x^2}}$$

****

use trig substitution:

$u^2 = x^2$ so $u = x$ $\quad 2\sin\theta = x \quad 2\cos\theta\, d\theta = dx$

$a^2 = 4$ so $a = 2$ $\quad 4\sin^2\theta = x^2$

$$\int \frac{3x^2\,dx}{\sqrt{4-x^2}} = 3\int \frac{4\sin^2\theta\; 2\cos\theta\, d\theta}{\sqrt{4-4\sin^2\theta}} = 24\int \frac{\sin^2\theta\cos\theta\, d\theta}{\sqrt{4(1-\sin^2\theta)}}$$

$$= 24\int \frac{\sin^2\theta\cos\theta\, d\theta}{2\cos\theta} = 12\int \sin^2\theta\, d\theta = 12\int \frac{1-\cos 2\theta}{2}\, d\theta$$

$$= 6\int (1-\cos 2\theta)\, d\theta = 6\left(\theta - \tfrac{1}{2}\sin 2\theta\right) + C$$

$$= 6\left(\theta - \tfrac{1}{2}(2)\sin\theta\cos\theta\right) + C = 6\theta - 6\sin\theta\cos\theta + C$$

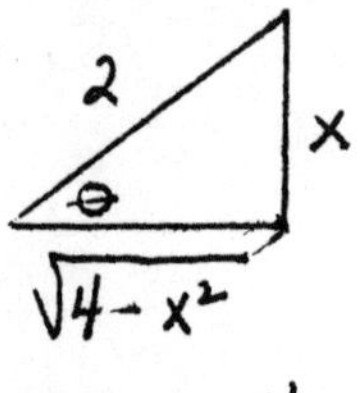

$$= 6\,\text{Arcsin}(x/2) - (6)(x/2)\left(\frac{\sqrt{4-x^2}}{2}\right) + C$$

$$= 6\,\text{Arcsin}(x/2) - (3/2)\,x\sqrt{4-x^2} + C$$

$\sin\theta = x/2$

$\cos\theta = \dfrac{\sqrt{4-x^2}}{2}$

$\theta = \text{Arcsin}(x/2)$

2-40

Evaluate the following indefinite integral

$$\int \frac{x^3 + 4x^2 + 13x + 3}{x^2 + 4x + 13}\, dx$$

****

Since the degree of the numerator is larger than the degree of the denominator, divide the numerator by the denominator and obtain

$$\frac{x^3+4x^2+13x+3}{x^2+4x+13} = x + \frac{3}{x^2+4x+13}$$

$$\therefore \int \frac{x^3+4x^2+13x+3}{x^2+4x+13}\, dx = \int x\, dx + 3\int \frac{dx}{x^2+4x+13}$$

$$= \frac{x^2}{2} + 3\int \frac{dx}{(x^2+4x+4)+9} = \frac{x^2}{2} + 3\int \frac{dx}{(x+2)^2+9}$$

$$= \frac{x^2}{2} + 3\cdot\frac{1}{3}\tan^{-1}\left(\frac{x+2}{3}\right) + C$$

$$= \tfrac{1}{2}x^2 + \tan^{-1}\left(\frac{x+2}{3}\right) + C$$

2-41

If an integral involves the quantity $\sqrt{x^2 + 1}$, a good substitution to try is (a) $x = \sin u$ (b) $x = \tan u$ (c) $x = \sec u$ (d) $x = e^u$ (e) $x = \ln u$.

****

$x = \tan u$ will eliminate the radical since $\sqrt{x^2+1} =$

$\sqrt{\tan^2 u + 1} = \sqrt{\sec^2 u} = \sec u$.

2-42

Integrate:

$$\int \frac{\sqrt{x^2-4}}{x}\,dx$$

****

Let $x = 2\sec\theta$

$\Rightarrow dx = 2\sec\theta\tan\theta\,d\theta$

$$\int \frac{\sqrt{x^2-4}}{x}\,dx = \int \frac{\sqrt{4\sec^2\theta-4}}{\cancel{2\sec\theta}}\,\cancel{2\sec\theta}\tan\theta\,d\theta$$

$$= \int \sqrt{4(\sec^2\theta-1)}\,\tan\theta\,d\theta$$

$$= 2\int(\tan\theta)(\tan\theta)\,d\theta = 2\int \tan^2\theta\,d\theta$$

$$= 2\int(\sec^2\theta-1)\,d\theta = 2[\tan\theta-\theta]+C$$

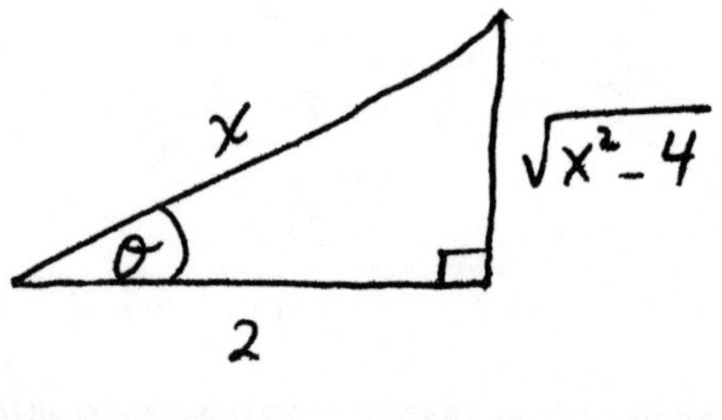

$$= 2\left[\frac{\sqrt{x^2-4}}{2} - \text{arc sec}\,\frac{x}{2}\right]+C$$

$$= \sqrt{x^2-4} - 2\,\text{arc sec}\left(\frac{x}{2}\right)+C$$

2-43

Use a trigonometric substitution to evaluate

$$\int_0^2 \frac{x^2}{(4 + x^2)^2}\, dx$$

****

Let $x = 2\tan\theta$ then $\frac{dx}{d\theta} = 2\sec^2\theta \Rightarrow dx = 2\sec^2\theta\, d\theta$

Now $\theta = \tan^{-1}(x/2)$ so when $x = 2$, $\theta = \pi/4$
and when $x = 0$, $\theta = 0$

So

$$\int_0^2 \frac{x^2}{(4+x^2)^2}\, dx = \int_0^{\pi/4} \frac{4\tan^2\theta}{(4 + 4\tan^2\theta)^2} \cdot 2\sec^2\theta\, d\theta$$

$$= \int_0^{\pi/4} \frac{8\tan^2\theta \sec^2\theta}{16\sec^4\theta}\, d\theta$$

$$= \frac{1}{2}\int_0^{\pi/4} \frac{\tan^2\theta}{\sec^2\theta}\, d\theta$$

$$= \frac{1}{2}\int_0^{\pi/4} \sin^2\theta\, d\theta$$

$$= \frac{1}{4}\int_0^{\pi/4} (1 - \cos 2\theta)\, d\theta$$

$$= \frac{1}{4}\left(\theta - \frac{\sin 2\theta}{2}\right)\Bigg]_0^{\pi/4}$$

$$= \frac{\pi - 2}{16}$$

2-44

Evaluate the following: $\int \sqrt{1-x^2}\, dx$

****

Trig Substitution

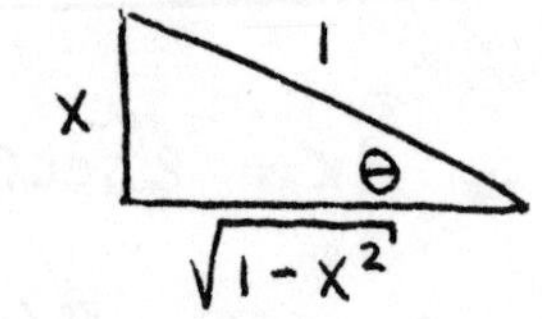

$\sin\theta = x \qquad dx = \cos\theta\, d\theta$

$\cos\theta = \sqrt{1-x^2}$

$$\int \sqrt{1-x^2}\, dx = \int (\cos\theta)\cos\theta\, d\theta = \int \cos^2\theta\, d\theta$$

$$= \int \frac{1+\cos 2\theta}{2}\, d\theta = \frac{1}{2}\left(\theta + \frac{\sin 2\theta}{2}\right) + C$$

$$= \frac{1}{2}\left(\theta + \frac{2\sin\theta\cos\theta}{2}\right) + C = \frac{1}{2}\left(\sin^{-1}x + x\sqrt{1-x^2}\right) + C$$

$$= \frac{1}{2}\sin^{-1}x + \frac{1}{2}x\sqrt{1-x^2} + C$$

2-45

Find $\int \frac{1}{x(x^2+1)^{3/2}}\, dx$.

****

$$\int \frac{1}{x(x^2+1)^{3/2}}\, dx = \int \frac{\sec^2 u\, du}{\tan u \sec^3 u} = \int \frac{du}{\tan u \sec u} = \int \frac{\cos^2 u}{\sin u}\, du$$

let $x = \tan u$

$dx = \sec^2 u\, du$

$$= \int \frac{1-\sin^2 u}{\sin u}\, du = \int (\csc u - \sin u)\, du$$

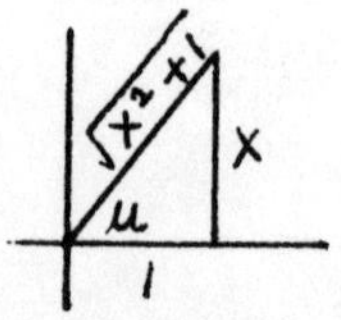

$$= \ln|\csc u - \cot u| + \cos u + C$$

$$= \ln\left|\frac{\sqrt{x^2+1}}{x} - \frac{1}{x}\right| + \frac{1}{\sqrt{x^2+1}} + C.$$

2-46

Find $\int \sqrt{x^2 + 5}\ dx$

****

$$\int \sqrt{x^2+5}\, dx$$

$$= \sqrt{5}\int \sqrt{\left(\frac{x}{\sqrt{5}}\right)^2+1}\ dx \qquad \text{LET } \frac{x}{\sqrt{5}} = \tan u$$

$$dx = \sqrt{5}\sec^2 u\, du$$

$$= \sqrt{5}\int \sqrt{\tan^2 u+1}\ \sqrt{5}\sec^2 u\, du$$

$$= 5\int \sec^3 u\, du \qquad \text{LET } w = \sec u$$

$$dw = \sec u \tan u\, du$$

$$= 5\left[\sec u \tan u - \int \sec u \tan^2 u\, du\right] \qquad dz = \sec^2 u\, du$$

$$= 5\sec u \tan u - 5\int \sec u(\sec^2 u - 1)\, du \qquad z = \tan u$$

$$= 5\sec u \tan u - 5\left[\int \sec^3 u\, du - \int \sec u\, du\right] \qquad \text{(BY PARTS)}$$

NOW, $5\int \sec^3 u\, du = 5\sec u \tan u - 5\int \sec^3 u\, du + 5\int \sec u\, du$

SO THAT $10\int \sec^3 u\, du = 5\sec u \tan u + 5\int \sec u\, du$

$$5\int \sec^3 u\, du = \frac{5}{2}\sec u \tan u + \frac{5}{2}\log\left|\sec u + \tan u\right| + C$$

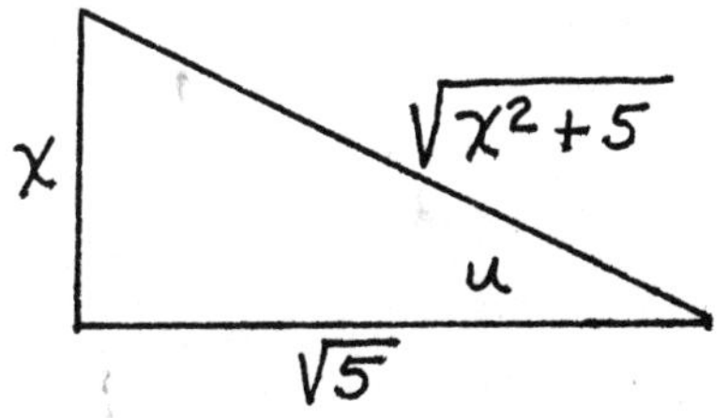

$$\int \sqrt{x^2+5}\, dx = \frac{5}{2}\cdot\frac{\sqrt{x^2+5}}{\sqrt{5}}\cdot\frac{x}{\sqrt{5}} + \frac{5}{2}\log\left|\frac{\sqrt{x^2+5}}{\sqrt{5}} + \frac{x}{\sqrt{5}}\right| + C$$

$$= \frac{x\sqrt{x^2+5}}{2} + \frac{5}{2}\log\left(\sqrt{x^2+5} + x\right) + C$$

2-47

Find $\displaystyle\int \frac{dx}{\sqrt{x^2 + 2x + 2}}$

****

Letting $u = x+1$,

$$\int \frac{dx}{\sqrt{x^2+2x+2}} = \int \frac{dx}{\sqrt{(x+1)^2+1}} = \int \frac{du}{\sqrt{u^2+1}}$$

Now substitute $u = \tan\theta$ $(du = \sec^2\theta \, d\theta)$

$$= \int \frac{\sec^2\theta \, d\theta}{\sqrt{\tan^2\theta + 1}} = \int \frac{\sec^2\theta \, d\theta}{\sqrt{\sec^2\theta}} = \int \sec\theta \, d\theta$$

$$= \ln\left|\sec\theta + \tan\theta\right| + c$$

$$= \ln\left|\sqrt{u^2+1} + u\right| + c$$

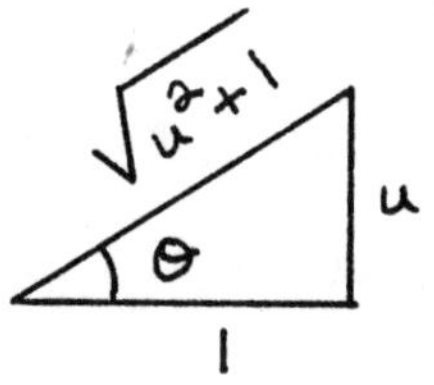

$$= \ln\left|\sqrt{x^2+2x+2} + (x+1)\right| + c$$

2-48

Evaluate the following integral

$$\int \frac{dx}{2x\sqrt{9 - x^2}}$$

****

Let $x = 3\sin(\theta)$

$dx = 3\cos(\theta)\, d\theta$

$\sqrt{9-x^2} = \sqrt{9-9\sin^2(\theta)} = 3\sqrt{1-\sin^2(\theta)} = 3\sqrt{\cos^2(\theta)}$

$= 3\cos(\theta)$

Then

$$\int \frac{dx}{2x\sqrt{9-x^2}} = \frac{1}{2}\int \frac{3\cos(\theta)\, d\theta}{3\sin(\theta)\cdot 3\cos(\theta)} = \frac{1}{6}\int \csc(\theta)\, d\theta$$

$$= \frac{1}{6}\ln\left|\csc(\theta) - \cot(\theta)\right| + C$$

Since $x = 3\sin(\theta)$, $\sin(\theta) = \frac{x}{3}$. Using a right triangle, it can be seen that

$\csc(\theta) = \frac{3}{x}$ and $\cot(\theta) = \frac{\sqrt{9-x^2}}{x}$

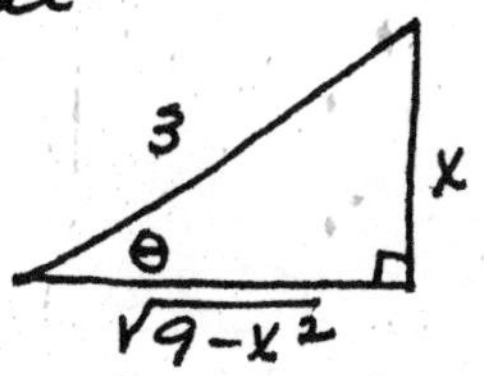

So

$$\int \frac{dx}{2x\sqrt{9-x^2}} = \frac{1}{6}\ln\left|\frac{3}{x} - \frac{\sqrt{9-x^2}}{x}\right| + C$$

$$= \frac{1}{6}\ln\left|\frac{3-\sqrt{9-x^2}}{x}\right| + C$$

INTEGRATION OF RATIONAL FUNCTIONS OF SINE AND COSINE

2-49

Integrate

$$\int \frac{\sin x \, dx}{\cos x + \cos^2 x}$$

****

$$\text{Let } I = \int \frac{\sin x \, dx}{\cos x + \cos^2 x} = \int \frac{\tan x \, dx}{1 + \cos x}$$

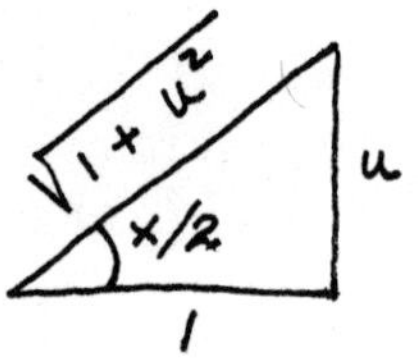

$$u = \tan \frac{x}{2} \qquad du = \frac{1}{2} \sec^2 \frac{x}{2} \, dx$$

$$\frac{u}{\sqrt{1+u^2}} = \sin \frac{x}{2} \qquad du = \frac{1}{2}(1 + u^2) \, dx$$

$$dx = \frac{2 \, du}{1 + u^2}$$

$$\frac{1}{\sqrt{1+u^2}} = \cos \frac{x}{2}$$

$$\tan x = \frac{2 \tan \frac{x}{2}}{1 - \tan^2 \frac{x}{2}} = \frac{2u}{1 - u^2} \qquad \cos x = \cos^2 \frac{x}{2} - \sin^2 \frac{x}{2}$$

$$\cos x = \frac{1 - u^2}{1 + u^2}$$

$$I = \int \frac{\frac{2u}{1 - u^2} \, \frac{2 \, du}{1 + u^2}}{1 + \frac{1 - u^2}{1 + u^2}} = \int \frac{4u \, du}{(1 - u^2)(1 + u^2 + 1 - u^2)} = \int \frac{2u \, du}{1 - u^2}$$

$$= -\ln \left| 1 - u^2 \right| + c = -\ln \left| 1 - \tan^2 \frac{x}{2} \right| + c$$

2-50

Evaluate the following indefinite integral

$$\int \frac{dx}{\sin(x) + \cos(x)}$$

****

Let $z = \tan(\frac{1}{2}x)$. Using the identities $\cos(2y) = 2\cos^2(y) - 1$ and $y = \frac{1}{2}x$ it can be shown that

$$\cos(x) = \frac{1-z^2}{1+z^2}$$

Using the identities $\sin(2y) = 2\sin(y)\cos(y)$ and $y = \frac{1}{2}x$ it can be shown that

$$\sin(x) = \frac{2z}{1+z^2}.$$

Also $d(z) = d[\tan(\frac{1}{2}x)] = \frac{1}{2}\sec^2(\frac{1}{2}x)\,dx$

$$= \tfrac{1}{2}[1+\tan^2(\tfrac{1}{2}x)]dx = \tfrac{1}{2}(1+z^2)\,dx.$$

So $dx = \frac{2dz}{1+z^2}$. Substituting these into the integral yields

$$\int \frac{\frac{2dz}{1+z^2}}{\frac{2z}{1+z^2} + \frac{1-z^2}{1+z^2}} = \int \frac{2dz}{1+2z-z^2} = 2\int \frac{dz}{1-(z^2-2z)}$$

$$= 2\int \frac{dz}{2-(z^2-2z+1)} = 2\int \frac{dz}{2-(z-1)^2}$$

Since $\int \frac{du}{a^2-u^2} = \frac{1}{2a}\ln\left|\frac{u+a}{u-a}\right| + C$,

$$\int \frac{dx}{\sin(x)+\cos(x)} = 2 \cdot \frac{1}{2\sqrt{2}} \ln\left|\frac{z-1+\sqrt{2}}{z-1-\sqrt{2}}\right| + C$$

$$= \frac{1}{\sqrt{2}} \ln\left|\frac{\tan(\frac{1}{2}x) - 1 + \sqrt{2}}{\tan(\frac{1}{2}x) - 1 - \sqrt{2}}\right| + C$$

2-51

Use $z = \tan(x/2)$ to find $\int \frac{\cos x\, dx}{1 + \cos x}$.

****

$$\int \frac{\cos x\, dx}{1+\cos x} = \int \frac{\frac{1-z^2}{1+z^2}\ \frac{2dz}{1+z^2}}{1+\frac{1-z^2}{1+z^2}}\, dz = \int \frac{\frac{2(1-z^2)\,dz}{(1+z^2)^2}}{\frac{2}{1+z^2}}$$

$$= \int \frac{1-z^2}{1+z^2}\,dz \underset{\text{by long division}}{=} \int \left[-1 + \frac{2}{z^2+1}\right] dz = -z + 2\tan^{-1} z + C$$

$$= -\tan\frac{x}{2} + x + C.$$

INTEGRALS YIELDING INVERSE HYPERBOLIC FUNCTIONS

2-52

Use the substitution $x = \sinh u$ to integrate the following.

$$\int \sqrt{1 + x^2}\, dx$$

****

Let $\mathcal{I} = \int \sqrt{1+x^2}\, dx$ and $x = \sinh u$

Then $\mathcal{I} = \int \sqrt{1+\sinh^2 u}\, \cosh u\, du$

$= \int \cosh u \cosh u\, du$

$= \int \cosh^2 u\, du = \frac{1}{2}\int (\cosh 2u + 1)\, du$

$= \frac{1}{2}\left(\frac{1}{2}\sinh 2u + u\right) + c$

$= \frac{1}{2}(\sinh u \cosh u + u) + c$

$= \frac{1}{2}\left(x\sqrt{1+x^2} + \operatorname{argsinh} x\right) + c$

2-53

Evaluate the integral $\int \frac{dx}{\sqrt{x^2 - 10x + 29}}$ by using a hyperbolic function substitution and then check the result with a formula.

****

We first transform the denominator in the form of a perfect square by writing:

$$\int \frac{dx}{\sqrt{x^2 - 10x + 29}} = \int \frac{dx}{\sqrt{x^2 - 10x + 25 + 4}}$$

$$= \int \frac{dx}{\sqrt{(x-5)^2 + 4}}$$

Now, let $(x - 5) = 2 \sinh u$ .....(1)

so that $\frac{dx}{du} = 2 \cosh u$

$dx = 2 \cosh u \cdot du$

Also, by rearranging Eq.(1) we have

$$u = \sinh^{-1} \frac{1}{2}(x - 5)$$

Substituting the calculated values into the rearranged problem statement we have

$$\int \frac{dx}{\sqrt{(x-5)^2+4}} = \int \frac{2 \cosh u \, du}{\sqrt{(2 \sinh u)^2 + 4}}$$

$$= \int \frac{2 \cosh u \, du}{\sqrt{4 \sinh^2 u + 4}}$$

$$= \int \frac{2 \cosh u \, du}{2\sqrt{\sinh^2 u + 1}}$$

By using the identity $\cosh^2 t - \sinh^2 t = 1$ we write

$$\int \frac{2 \cosh u \, du}{2\sqrt{\sinh^2 u + 1}} = \int \frac{2 \cosh u \, du}{2\sqrt{\cosh^2 u}}$$

$$= \int \frac{2 \cosh u \, du}{2 \cosh u}$$

$$= \int du$$

$$= u + C$$

$$= \sinh^{-1}\left(\frac{x-5}{2}\right) + C$$

which indeed checks with the formula

$$\int \frac{du}{\sqrt{u^2 + a^2}} = \sinh^{-1} \frac{u}{a} + C$$

2-54

Express the following indefinite integral in terms of an inverse hyperbolic function and as a natural logarithm:

$$\int \frac{t\,dt}{\sqrt{t^4 + 12t^2 + 40}}$$

****

$$\int \frac{t\,dt}{\sqrt{t^4+12t^2+40}} = \int \frac{t\,dt}{\sqrt{(t^4+12t^2+36)+4}} = \int \frac{t\,dt}{\sqrt{(t^2+6)^2+4}}$$

Let $u = t^2+6$, then $du = 2t\,dt$ or $t\,dt = \frac{du}{2}$

$$\int \frac{t\,dt}{\sqrt{t^4+12t^2+40}} = \frac{1}{2}\int \frac{du}{\sqrt{u^2+4}}$$

$$= \frac{1}{2}\sinh^{-1}\left(\frac{u}{2}\right) + C$$

$$= \frac{1}{2}\sinh^{-1}\left(\frac{t^2+6}{2}\right) + C$$

and

$$= \ln\left(u + \sqrt{u^2+a^2}\right) + C$$

$$= \ln\left(t^2+6+\sqrt{t^4+12t^2+40}\right) + C$$

DECOMPOSITION OF RATIONAL FUNCTIONS INTO PARTIAL FRACTIONS

2-55

Evaluate the integral $\int \frac{5x^2 + 26x + 29}{(x+2)(x+1)(x+3)}\, dx.$

****

LET $\frac{5x^2+26x+29}{(x+2)(x+1)(x+3)} = \frac{A}{x+2} + \frac{B}{x+1} + \frac{C}{x+3}$

$$= \frac{A(x+1)(x+3)+B(x+2)(x+3)+C(x+2)(x+1)}{(x+2)(x+1)(x+3)}$$

SO $A(x^2+4x+3) + B(x^2+5x+6) + C(x^2+3x+2) = 5x^2+26x+29$

$(A+B+C)\,x^2 + (4A+5B+3C)x + (3A+6B+2C) = 5x^2+26x+29$

THEREFORE,
$$A + B + C = 5$$
$$4A + 5B + 3C = 26$$
$$3A + 6B + 2C = 29$$

$$\begin{array}{r} -3A - 3B - 3C = -15 \\ 4A + 5B + 3C = 26 \\ \hline A + 2B \quad = 11 \end{array} \qquad \begin{array}{r} -2A - 2B - 2C = -10 \\ 3A + 6B + 2C = 29 \\ \hline A + 4B \quad = 19 \end{array}$$

$$\begin{array}{r} -2A - 4B = -22 \\ A + 4B = 19 \\ \hline -A \quad = -3 \\ A = 3 \end{array}$$

SINCE $A = 3$, $\quad 3 + 4B = 19$, $\quad 3 + 4 + C = 5$.
$$B = 4 \qquad C = -2$$

THEREFORE, $\int \frac{5x^2+26x+29}{(x+2)(x+1)(x+3)}dx = \int \left(\frac{3}{x+2} + \frac{4}{x+1} + \frac{-2}{x+3}\right)dx$

$$= 3\ln|x+2| + 4\ln|x+1| - 2\ln|x+3| + C.$$

2-56

Integrate $\displaystyle\int \frac{(2x + 1)}{(x^2 + 1)(3x - 1)} \, dx$

****

Set $\displaystyle\frac{2x+1}{(x^2+1)(3x-1)} = \frac{Ax+B}{x^2+1} + \frac{C}{3x-1}$

Multiply both sides of the equation by $(x^2+1)(3x-1)$:

$$2x+1 = (Ax+B)(3x-1) + C(x^2+1)$$

$$= 3Ax^2 - Ax + 3Bx - B + Cx^2 + C$$

$$= (3A+C)x^2 + (3B-A)x + C - B$$

Set coefficients of like functions equal:

$$3A+C = 0 \qquad 3B-A = 2 \qquad C-B = 1$$

Solve simultaneously: $A = -\frac{1}{2} \qquad B = \frac{1}{2} \qquad C = \frac{3}{2}$

$$\frac{2x+1}{(x^2+1)(3x-1)} = \frac{-\frac{1}{2}x + \frac{1}{2}}{x^2+1} + \frac{\frac{3}{2}}{3x-1}$$

$$\int \frac{2x+1}{(x^2+1)(3x-1)} \, dx = \int \frac{-\frac{1}{2}x + \frac{1}{2}}{x^2+1} \, dx + \int \frac{\frac{3}{2}}{3x-1} \, dx$$

$$= -\frac{1}{2}\int \frac{x\,dx}{x^2+1} + \frac{1}{2}\int \frac{dx}{x^2+1} + \frac{3}{2}\int \frac{dx}{3x-1}$$

Let $u = x^2+1 \qquad du = 2x\,dx \qquad\qquad v = 3x-1 \qquad dv = 3dx$

$$= -\frac{1}{2}\int \frac{\frac{1}{2}du}{u} + \frac{1}{2}\int \frac{dx}{x^2+1} + \frac{3}{2}\int \frac{\frac{1}{3}dv}{v}$$

$$= -\frac{1}{4}\ln|u| + \frac{1}{2}\text{Arctan } x + \frac{1}{2}\ln|v| + C$$

$$= -\frac{1}{4}\ln(x^2+1) + \frac{1}{2}\text{Arctan } x + \frac{1}{2}\ln|3x-1| + C.$$

2-57

Evaluate the following indefinite integral

$$\int \frac{12 + 21x - 8x^2}{4x^2 - x^3}\,dx$$

****

The denominator factors as $x^2(4-x)$, two linear factors where one repeats. Decompose the rational function into partial fractions

$$\frac{12+21x-8x^2}{x^2(4-x)} = \frac{A}{x^2} + \frac{B}{x} + \frac{C}{4-x}$$

$$= \frac{A(4-x) + Bx(4-x) + Cx^2}{x^2(4-x)}$$

$\therefore$

$$12+21x-8x^2 = 4A - Ax + 4Bx - Bx^2 + Cx^2$$
$$= 4A + (4B-A)x + (C-B)x^2$$

This leads to the following system of equations:

$4A = 12 \longrightarrow A = 3$

$4B - A = 21$ then $4B - 3 = 21$, $4B = 24$, $B = 6$

$C - B = -8$ then $C - 6 = -8$, $C = -2$

So

$$\int \frac{12+21x-8x^2}{4x^2-x^3}\,dx = \int \frac{3}{x^2}\,dx + \int \frac{6}{x}\,dx + \int \frac{-2}{4-x}\,dx$$

$$= -\frac{3}{x} + 6\ln|x| + 2\ln|4-x| + C$$

$$= -\frac{3}{x} + 2\left[\ln|x^3| + \ln|4-x|\right] + C$$

$$= -\frac{3}{x} + 2\ln\left|x^3(4-x)\right| + C$$

2-58

Evaluate the following indefinite integral:

$$\int \frac{25}{x^4 + 2x^3 + 5x^2}\,dx$$

****

Using partial fractions,

$$\frac{25}{x^4+2x^3+5x^2} = \frac{25}{x^2(x^2+2x+5)} = \frac{A}{x} + \frac{B}{x^2} + \frac{Cx+D}{x^2+2x+5}$$

$$\begin{aligned} 25 &= Ax(x^2+2x+5) + B(x^2+2x+5) + (Cx+D)x^2 \\ &= Ax^3+2Ax^2+5Ax + Bx^2+2Bx+5B + Cx^3+Dx^2 \\ &= x^3(A+C) + x^2(2A+B+D) + x(5A+2B) + 5B \end{aligned}$$

$25 = 5B \Rightarrow B = 5$

$5A + 2B = 5A + 10 = 0 \Rightarrow A = -2$

$2A + B + D = -4 + 5 + D = 0 \Rightarrow D = -1$

$A + C = -2 + C = 0 \Rightarrow C = 2$

$$\therefore \int \frac{25}{x^4+2x^3+5x^2}\,dx = \int \frac{-2}{x}\,dx + \int \frac{5}{x^2}\,dx$$

$$+ \int \frac{2x-1}{x^2+2x+5}\,dx = -2\ln|x| - \frac{5}{x} + \int \frac{2x+2}{x^2+2x+5}\,dx$$

$$- 3\int \frac{1}{(x+1)^2+2^2}\,dx = -2\ln|x| - \frac{5}{x} + \ln(x^2+2x+5)$$

$$- \frac{3}{2}\tan^{-1}\frac{x+1}{2} + C$$

$$= \ln\left(\frac{x^2+2x+5}{x^2}\right) - \frac{5}{x} - \frac{3}{2}\tan^{-1}\frac{x+1}{2} + C$$

2-59

Perform the following integration:

$$\int \frac{(2x + 1)\ dx}{(x^2 - 1)(x^2 + 1)}$$

****

use the method of partial fractions.

find the decomposition

$$\frac{2x+1}{(x+1)(x-1)(x^2+1)} = \frac{A}{(x+1)} + \frac{B}{(x-1)} + \frac{Cx+D}{(x^2+1)}$$

$$\begin{aligned} 2x+1 &= A(x-1)(x^2+1) + B(x+1)(x^2+1) + (Cx+D)(x+1)(x-1) \\ &= Ax^3 - Ax^2 + Ax - A + Bx^3 + Bx^2 + Bx + B + Cx^3 + Dx^2 - Cx - D \\ &= (A+B+C)x^3 + (A-B+D)x^2 + (A+B-C)x + (A-B-D) \end{aligned}$$

matching coefficients on both sides of equation

① $0 = A+B+C$ add eqn 1 + eqn 3: $2A + 2B = 2$

② $0 = A-B+D$ add eqn 2 + eqn 4: $2A - 2B = 1$

③ $2 = A+B-C$ $4A = 3 \Rightarrow A = 3/4$

④ $1 = A-B-D$

$B = 1/4$

$C = -1$

$D = -1/2$

so:

$$\int \frac{2x+1}{(x^2-1)(x^2+1)}\,dx = \int \frac{3/4}{x+1} + \frac{1/4}{x-1} + \frac{-x-1/2}{x^2+1}\,dx$$

$$= \int \frac{3/4}{x+1} + \frac{1/4}{x-1} - \frac{x}{x^2+1} - \frac{1/2}{x^2+1}\,dx$$

$$= 3/4 \ln|x+1| + 1/4 \ln|x-1| - \tfrac{1}{2}\ln|x^2+1| - \tfrac{1}{2}\arctan x + C$$

2-60

Find $\int \frac{1}{x^3-x^2+x-1}\,dx$ using partial fractions.

****

Now $\frac{1}{x^3-x^2+x-1} = \frac{1}{(x^2+1)(x-1)}$

By partial fractions $\frac{1}{(x^2+1)(x-1)} = \frac{Ax+B}{x^2+1} + \frac{C}{x-1}$

So

$$1 = (Ax+B)(x-1) + C(x^2+1).$$

If $x=0$, $B(-1)+C=1$
If $x=1$, $C(1+1) = 1$

Using the above Equations we get
$C=\frac{1}{2}$ and $B=-\frac{1}{2}$

Since $B=-\frac{1}{2}$ and $C=\frac{1}{2}$ we may let $x=-1$ to get

$$A(-1+(-\tfrac{1}{2}))(-1-1) + \tfrac{1}{2}(2) = 1$$

So $A = 0$

Therefore $\frac{1}{x^3-x^2+x-1} = \frac{-\frac{1}{2}}{x^2+1} + \frac{\frac{1}{2}}{x-1}$

So $$\int \frac{1}{x^3-x^2+x-1}\,dx = \int\left(\frac{-\frac{1}{2}}{x^2+1} + \frac{\frac{1}{2}}{x-1}\right)dx$$

$$= -\frac{1}{2}\int\frac{1}{x^2+1}\,dx + \frac{1}{2}\int\frac{1}{x-1}\,dx$$

$$= -\frac{1}{2}\arctan x + \frac{1}{2}\ln(x-1) + C$$

2-61

Use the method of partial fractions to evaluate

$$\int_0^1 \frac{3x + 4}{x^3 - 2x - 4}\,dx$$

****

Now $x^3 - 2x - 4 = (x-2)(x^2 + 2x + 2)$

So $\dfrac{3x+4}{x^3 - 2x - 4} = \dfrac{A}{x-2} + \dfrac{Bx + C}{x^2 + 2x + 2}$

$\Rightarrow \quad 3x + 4 = A(x^2 + 2x + 2) + (Bx + C)(x - 2)$

When $x = 2$, $10 = A(10) \Rightarrow A = 1$

So $3x + 4 = (x^2 + 2x + 2) + (Bx^2 - 2Bx + Cx - 2C)$

$\Rightarrow \quad 3x + 4 = (1 + B)x^2 + (2 - 2B + C)x + (2 - 2C)$

Equating coefficients gives: $B = -1$ and $C = -1$

Thus, $\displaystyle\int_0^1 \frac{3x+4}{x^3 - 2x - 4}\,dx = \int_0^1 \left[\frac{1}{x-2} + \frac{(-x-1)}{x^2+2x+2}\right] dx$

$$= \left(\ln|x-2| - \frac{1}{2}\ln|x^2 + 2x + 2|\right)\Big]_0^1$$

$$= \frac{-1}{2}\ln(5) - \frac{1}{2}\ln(2)$$

$$= \left(\frac{-1}{2}\right)\ln(10)$$

$$= \ln 10^{-1/2}$$

2-62

Integrate: $\displaystyle\int \frac{2x^2 + 5x - 4}{x(x+2)(x-1)}\,dx$

****

Use the method of partial fractions to rewrite the integrand as a sum. Set up the following identity:
(This is the case of linear distinct factors.)

$$\frac{2x^2+5x-4}{x(x+2)(x-1)} = \frac{A}{x} + \frac{B}{x+2} + \frac{C}{x-1},$$ where A, B and C are to be found.

Clearing fractions yields:

$$2x^2+5x-4 = A(x+2)(x-1) + Bx(x-1) + Cx(x+2).$$ Now substitute:

$x = -2$, $8-10-4 = 6B$ and $B = -1$

$x = 0$, $-4 = A(2)(-1)$ and $A = 2$

$x = 1$, $2+5-4 = C(1)(3)$ and $C = 1$. Hence the original integral takes the form

$$\int \frac{2}{x}\,dx - \int \frac{dx}{x+2} + \int \frac{dx}{x-1} = 2\ln|x| - \ln|x+2| + \ln|x-1| + C.$$

NOTE: You may write the ans. in different form using the properties of logarithms.

2-63

Find $\displaystyle\int \frac{6x^3 + 3x + 1}{x^2(x^2 + 1)}\,dx$

****

$$\frac{6x^3+3x+1}{x^2(x^2+1)} = \frac{A}{x} + \frac{B}{x^2} + \frac{Cx+D}{x^2+1} \Rightarrow 6x^3+3x+1 = Ax(x^2+1) + B(x^2+1) + (Cx+D)x^2.$$

Letting $x=0$ gives $B=1$. Matching coefficients of x gives $A=3$. Matching coefficients of x^3 gives $A+C=6$, so $C=3$. Matching coefficients of x^2 gives $B+D=0$, so $D=-1$.

Therefore, $$\int \frac{6x^3+3x+1}{x^2(x^2+1)}\,dx = \int \left[\frac{3}{x} + \frac{1}{x^2} + \frac{3x}{x^2+1} - \frac{1}{x^2+1}\right] dx$$
$$= 3\ln|x| - \frac{1}{x} + \frac{3}{2}\ln(x^2+1) - \tan^{-1}x + C.$$

2-64

Find $\int \frac{x^2 + x + 4}{(x + 1)(x^2 + 3)} dx$

****

expressing the equation in terms of partial fractions

$$\frac{x^2+x+4}{(x+1)(x^2+3)} \equiv \frac{A}{(x+1)} + \frac{Bx+C}{x^2+3}$$

so $\frac{x^2+x+4}{(x+1)(x^2+3)} \equiv \frac{A(x^2+3)+(Bx+C)(x+1)}{(x+1)(x^2+3)}$

equating numerators $x^2+x+4 \equiv A(x^2+3)+(Bx+C)(x+1)$

To find the constants A, B and C, pick 3 values for x and equate both sides

(i) let $x=1$ say $\quad 6 = 4A+2B+2C$ — ①

(ii) let $x=0$ $\quad 4 = 3A+C$ — ②

(iii) let $x=-1$ $\quad 4 = 4A$ $\quad$ so $A=1$ hence from ② $C=1$ and from ① $B=0$

Thus $\int \frac{x^2+x+4}{(x+1)(x^2+3)} dx = \int \frac{1}{x+1} dx + \int \frac{1}{x^2+3} dx$

$$\int \frac{x^2+x+4}{(x+1)(x^2+3)} dx = \ln|x+1| + \frac{1}{\sqrt{3}} \arctan \frac{x}{\sqrt{3}} + C$$

2-65

Evaluate $\displaystyle\int \frac{dx}{x^3 + 2x^2 + x}$

****

$$\frac{1}{x^3+2x^2+x} = \frac{1}{x(x+1)^2} = \frac{A}{x} + \frac{B}{x+1} + \frac{C}{(x+1)^2}$$

$$= \frac{A(x+1)^2 + Bx(x+1) + Cx}{x(x+1)^2}$$

$$= \frac{(A+B)x^2 + (2A+B+C)x + A}{x(x+1)^2}$$

Equating coefficients,

$A+B = 0 = 2A+B+C$, $A = 1$. Thus

$B = C = -1$, and

$$\int \frac{dx}{x^3+2x^2+x} = \int \frac{1}{x} - \frac{1}{x+1} - \frac{1}{(x+1)^2}\, dx$$

$$= \ln|x| - \ln|x+1| + \frac{1}{x+1} + C$$

$$= \ln\left|\frac{x}{x+1}\right| + \frac{1}{x+1} + C$$

2-66

Find: $\int \frac{x\,dx}{(x-2)(x+3)}$

****

Using partial fraction decomposition, we have:

$$\frac{x}{(x-2)(x+3)} = \frac{A}{x-2} + \frac{B}{x+3}$$

So, $x = (x+3)A + (x-2)B$ for all x

using $x=-3$: $-3 = -5B$ or $B = 3/5$

using $x = 2$: $2 = 5A$ or $A = 2/5$

hence: $$\frac{x}{(x-2)(x+3)} = \frac{2/5}{x-2} + \frac{3/5}{x+3}$$

$$\int \frac{2}{5} \cdot \frac{dx}{x-2} = \frac{2}{5} \ln|x-2| + C$$

$$\int \frac{3}{5} \cdot \frac{dx}{x+3} = \frac{3}{5} \ln|x+3| + C$$

Hence, $$\int \frac{x}{(x-2)(x+3)}\,dx = \frac{2}{5} \ln|x-2| + \frac{3}{5} \ln|x+3| + C$$

2-67

Evaluate the integral $\int \frac{2x\,dx}{(x^2+1)(x+1)^2}$

****

$$\frac{2x}{(x^2+1)(x+1)^2} = \frac{Ax+B}{x^2+1} + \frac{C}{x+1} + \frac{D}{(x+1)^2}$$

$$2x = (Ax+B)(x+1)^2 + C(x+1)(x^2+1) + D(x^2+1)$$

$$2x = Ax^3+2Ax^2+Ax+Bx^2+2Bx+B+Cx^3+Cx^2+Cx+C+Dx^2+D$$

$$2x = (A+C)x^3+(2A+B+C+D)x^2+(A+2B+C)x + B+C+D$$

$$\therefore \begin{cases} A+C=0 \\ 2A+B+C+D=0 \\ A+2B+C=2 \\ B+C+D=0 \end{cases} \Longrightarrow \begin{array}{l} A=0 \\ B=1 \\ C=0 \\ D=-1 \end{array}$$

$$\therefore \int \frac{2x\,dx}{(x^2+1)(x+1)^2} = \int \left(\frac{1}{x^2+1} - \frac{1}{(x+1)^2}\right)dx$$

$$= \arctan x + \frac{1}{x+1} + C$$

MISCELLANEOUS SUBSTITUTIONS

2-68

Find $\int \frac{\sqrt{x+2} - 1}{\sqrt{x+2} + 1}\, dx$

****

LET $x+2 = u^2$

$dx = 2u\,du$

$$\int \frac{\sqrt{x+2} - 1}{\sqrt{x+2} + 1}\, dx$$

$$= \int \frac{u-1}{u+1}(2u\,du)$$

$$= 2\int \frac{u^2-u}{u+1}\, du$$

$$= 2\int \left(u - 2 + \frac{2}{u+1}\right) du$$

$$= u^2 - 4u + 4\log|u+1| + C$$

$$= x + 2 - 4\sqrt{x+2} + 4\log\left|\sqrt{x+2} + 1\right| + C$$

2-69

Perform the following integration:

$$\int \frac{2 + \sqrt[3]{x}}{\sqrt[3]{x} + \sqrt{x}}\, dx$$

****

let $u = x^{1/6}$ so: $\int \frac{2+\sqrt[3]{x}}{\sqrt[3]{x}+\sqrt{x}}\,dx = \int \frac{(2+u^2)\,6u^5\,du}{u^2+u^3}$

$u^2 = x^{1/3}$

$u^3 = x^{1/2}$

$u^6 = x$ $\quad = 6\int \frac{(2+u^2)u^3\,du}{1+u} = 6\int \frac{u^5+2u^3\,du}{u+1}$

$6u^5\,du = dx$

by long division $= 6\int u^4 - u^3 + 3u^2 - 3u + 3 - 3/u+1\; du$

$= 6\left(u^5/5 - u^4/4 + 3u^3/3 - 3u^2/2 + 3u - 3\ln|u+1|\right) + C$

$= 6\left(\frac{1}{5}x^{5/6} - \frac{1}{4}x^{4/6} + x^{3/6} - \frac{3}{2}x^{2/6} + 3x^{1/6} - 3\ln|x^{1/6}+1|\right) + C$

$= \frac{6}{5}x^{5/6} - \frac{3}{2}x^{2/3} + 6x^{1/2} - 9x^{1/3} + 18x^{1/6} - 18\ln|x^{1/6}+1| + C$

2-70

Evaluate the integral $\int x^2\sqrt{x-2}\,dx$

****

LET $\sqrt{x-2} = u$

$\therefore\ x-2 = u^2$

HENCE, $dx = 2u\,du$

$$\therefore \int x^2\sqrt{x-2}\,dx = \int (u^2+2)^2 u\,(2u\,du)$$

$$= \int (2u^6+8u^4+8u^2)\,du$$

$$= \frac{2}{7}u^7 + \frac{8}{5}u^5 + \frac{8}{3}u^3 + C$$

$$= \frac{2}{7}(x-2)^{\frac{7}{2}} + \frac{8}{5}(x-2)^{\frac{5}{2}} + \frac{8}{3}(x-2)^{\frac{3}{2}} + C$$

2-71

The substitution $x = u^{12}$ changes $\int \frac{1}{\sqrt[3]{x}+\sqrt[4]{x}}\,dx$ into (a) $\int \frac{1}{u^4+u^3}\,du$

(b) $\int \frac{12u^8}{u+1}\,du$ (c) $\int \frac{u^9}{u+1}\,du$ (d) $\int \frac{1}{u^2+u}\,du$ (e) $\int \frac{12u}{u^2+1}\,du$.

****

For $x = u^{12}$, $dx = 12u^{11}du$, so $\int \frac{1}{\sqrt[3]{x}+\sqrt[4]{x}}\,dx = \int \frac{1}{u^4+u^3}\cdot 12u^{11}du$

$$= \int \frac{12u^8}{u+1}\,du.$$

2-72

Integrate: $\int \frac{\sqrt[6]{x}\ dx}{\sqrt{x} + \sqrt[3]{x}}$

****

Substitute $y = \sqrt[6]{x}$ (6 is the L.C.D. of all the index values.) Then Raise to all powers up to and including six: $y^2 = \sqrt[3]{x}$, $y^3 = \sqrt{x}$, $y^4 = x^{2/3}$, $y^5 = x^{5/6}$ and $y^6 = x$.

Now differentiating $y^6 = x$ implicitly with Respect to x yields $6y^5 \cdot y' = 1$ oR $\frac{dy}{dx} = \frac{1}{6y^5}$. Hence $dx = 6y^5\,dy$ and the oRiginal integRal takes the following foRm by sub.:

$$\int \frac{y\,(6y^5\,dy)}{y^3 + y^2} = 6\int \frac{y^6\,dy}{y^2(y+1)} = 6\int \frac{y^4\,dy}{y+1}$$

.(Indeed ouR substitution "Rationalized" ouR integRand.) Now "divide out" synthetically to yield:

$$\frac{y^4}{y+1} = y^3 - y^2 + y - 1 + \frac{1}{y+1}$$

and sub. yields

$$6\int \frac{y^4}{y+1}\,dy = 6\int\left(y^3 - y^2 + y - 1 + \frac{1}{y+1}\right)dy$$

oR

$$6\int \frac{y^4}{y+1}\,dy = 6\left(\frac{y^4}{4} - \frac{y^3}{3} + \frac{y^2}{2} - y + \ln|y+1|\right) + c$$

and by back sub. the final answeR takes the foRm:

$$6\left(\frac{x^{2/3}}{4} - \frac{\sqrt{x}}{3} + \frac{\sqrt[3]{x}}{2} - \sqrt[6]{x} + \ln(\sqrt[6]{x} + 1)\right) + c$$

oR

$$\frac{3x^{2/3}}{2} - 2\sqrt{x} + 3\sqrt[3]{x} - 6\sqrt[6]{x} + 6\ln(\sqrt[6]{x} + 1) + c.$$

2-73

Give the integration technique most likely to work for each integral. It is not necessary to do the integration.

1. $\int x \sin x \, dx$

2. $\int \frac{x+1}{x^2-4} \, dx$

3. $\int \frac{t}{\sqrt{t+2}} \, dt$

4. $\int \frac{x^3+x^2+1}{x-3} \, dx$

5. $\int \sqrt{1-x^2} \, dx$

6. $\int \ln(2x-3) \, dx$

****

1. Integration by parts. The two factors, x and $\sin x$, are dissimilar.

2. Partial fractions. The integrand is a rational function, and the denominator factors.

3. Substitution, such as $u^2 = t+2$, to remove the radical.

4. Long division. The fraction is improper, since the numerator is of higher degree than the denominator. After division, integrate the quotient and remainder using basic formulas.

5. Trigonometric substitution $x = \sin\theta$ to remove the radical. $x = \cos\theta$ would work as well.

6. Integration by parts with $u = \ln(2x-3)$ and $dv = dx$ to change the form of the integral.

2-74

Evaluate $\int \frac{x^5 dx}{\sqrt{1 + x^3}}$

****

Let $u = 1+x^3$, $du = 3x^2 dx$.

Then $x^5 dx = \frac{(u-1)du}{3}$,

$$\int \frac{x^5 dx}{\sqrt{1+x^3}} = \frac{1}{3}\int \frac{u-1}{\sqrt{u}} du$$

$$= \frac{1}{3}\int u^{\frac{1}{2}} - u^{-\frac{1}{2}} du$$

$$= \frac{2}{9}u^{3/2} - \frac{2}{3}u^{1/2} + C$$

$$= \frac{2}{9}(1+x^3)^{3/2} - \frac{2}{3}(1+x^3)^{1/2} + C$$

3 APPLICATIONS OF THE DEFINITE INTEGRAL

ARC LENGTH

3-1

Find the length of the arc of the curve $9y^2 = 4(x-1)^3$ from (1,0) to (5,16/3).

****

$$9y^2 = 4(x-1)^3$$
$$y^2 = \frac{4}{9}(x-1)^3$$
$$y = \frac{2}{3}(x-1)^{3/2} \quad \text{SINCE } y \geq 0 \text{ FROM } (1,0) \text{ TO } (5,\tfrac{16}{3})$$

$$\int_1^5 \sqrt{1+\left(\frac{dy}{dx}\right)^2}\,dx = \int_1^5 \sqrt{1+[(x-1)^{1/2}]^2}\,dx$$
$$= \int_1^5 \sqrt{x}\,dx$$
$$= \frac{2}{3}x^{3/2}\Big|_1^5$$
$$= \frac{2}{3}5^{\frac{3}{2}} - \frac{2}{3} = \frac{10\sqrt{5}-2}{3}$$

3-2

The position of a particle P(x,y) at time t is given by

$$x = \frac{1}{3}(2t+3)^{3/2} \quad \text{and} \quad y = \frac{t^2}{2} + t.$$

Find the arc length of the path the particle would travel from t = 0 to t = 3.

****

$$\frac{dx}{dt} = \frac{1}{3}\left(\frac{3}{2}\right)(2t+3)^{\frac{1}{2}}(2) = (2t+3)^{\frac{1}{2}}$$

$$\frac{dy}{dt} = t+1$$

$$\therefore s = \int_0^3 \sqrt{\left(\frac{dx}{dt}\right)^2 + \left(\frac{dy}{dt}\right)^2}\, dt$$

$$= \int_0^3 \sqrt{2t+3+t^2+2t+1}\, dt$$

$$= \int_0^3 \sqrt{t^2+4t+4}\, dt = \int_0^3 \sqrt{(t+2)^2}\, dt$$

$$= \int_0^3 (t+2)\,dt = \left[\frac{t^2}{2} + 2t\right]_0^3 = \frac{9}{2} + 6 = \frac{21}{2}$$

3-3

Find the arc length of the curve $y = 2x^{3/2}$ between $x = 0$ and $x = 3$.

****

$y' = 3x^{1/2}$, $(y')^2 = 9x$

Letting $u = 1 + 9x$ $(du = 9\,dx)$,

$$\text{arc length} = \int_0^3 \sqrt{1+9x}\,dx = \frac{1}{9}\int_1^{28} \sqrt{u}\,du$$

$$= \frac{2}{27}\,u^{3/2}\Big|_1^{28} = \frac{2}{27}\left(28^{3/2} - 1\right)$$

3-4

Find the length of the curve $y = (8/3)x^{3/2}$ from the point $(1,8/3)$ to the point $(4,64/3)$.

****

$$\text{Arc length } S = \int_{x_1}^{x_2} \sqrt{1+(y')^2}\,dx$$

$y' = \frac{8}{3}\cdot\frac{3}{2}\,x^{1/2} = 4x^{1/2}$, so $(y')^2 = 16x$

$$S = \int_1^4 \sqrt{1+16x}\,dx \qquad \begin{pmatrix} u = 1+16x \\ du = 16\,dx \end{pmatrix}$$

$$= \int u^{1/2}\,\frac{1}{16}\,du = \frac{1}{16}\,\frac{u^{3/2}}{3/2}\Big| = \frac{1}{24}\left[(1+16x)^{3/2}\right]_1^4$$

$$= \frac{1}{24}\left[65^{3/2} - 17^{3/2}\right]$$

3-5

Set up, but do not evaluate, the equations and/or integrals to find the perimeter of the region bounded by the curve $y = x^2 - 2x$ and the x-axis.

****

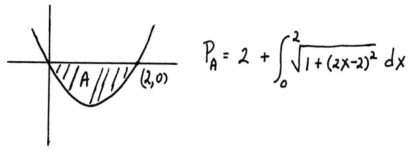

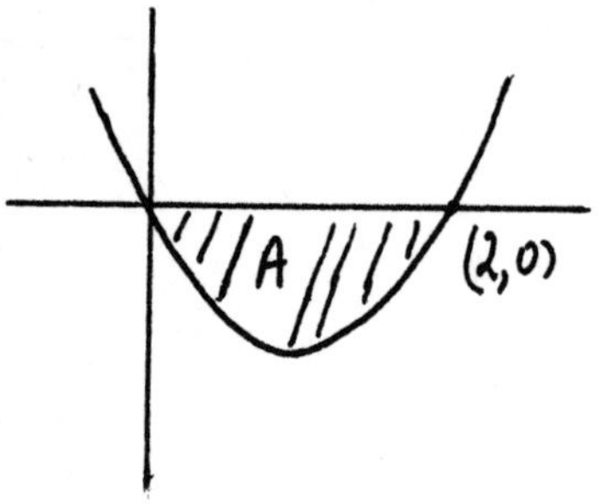

$$P_A = 2 + \int_0^2 \sqrt{1 + (2x-2)^2}\, dx$$

SURFACE AREA

3-6

Find the surface area when the graph of $f(x) = 3\sqrt{x}$, $x \in [0,2]$ is revolved about the x-axis.

****

$$f'(x) = \frac{3}{2} x^{-1/2}$$

$$\text{SURFACE AREA} = \int_0^2 2\pi (3\sqrt{x}) \sqrt{\left(\tfrac{3}{2}x^{-1/2}\right)^2 + 1}\, dx$$

$$= 3\pi \int_0^2 2\sqrt{x} \sqrt{\tfrac{9}{4}x^{-1} + 1}\, dx$$

$$= 3\pi \int_0^2 \sqrt{9 + 4x}\, dx$$

$$= \frac{3\pi}{4} \left[\frac{2}{3} (9 + 4x)^{3/2} \right]_0^2$$

$$= \frac{\pi}{2} \left(17^{3/2} - 27 \right)$$

3-7

Find the surface area generated when the quarter-circle $x^2 + y^2 = 4$ in the first octant is revolved around the y-axis.

****

$$S.A. = \int_0^2 2\pi x\sqrt{1+y'^2}\,dx$$

$$= \int_0^2 2\pi x\sqrt{1+\frac{x^2}{4-x^2}}\,dx$$

$$= \int_0^2 2\pi x\sqrt{\frac{4-x^2+x^2}{4-x^2}}\,dx$$

$$= 2\pi\int_0^2 \frac{2x}{\sqrt{4-x^2}}\,dx \text{. Let } u = 4-x^2,\ du = -2x\,dx$$

$$= 2\pi\int_4^0 \frac{-du}{\sqrt{u}}$$

$$= 2\pi \cdot 2\sqrt{u}\Big|_0^4 = 8\pi.$$

3-8

Prove that the lateral area of a frustum of a right circular cone with radii r_1 and r_2 and slant height s is $\pi(r_1 + r_2)s$.

****

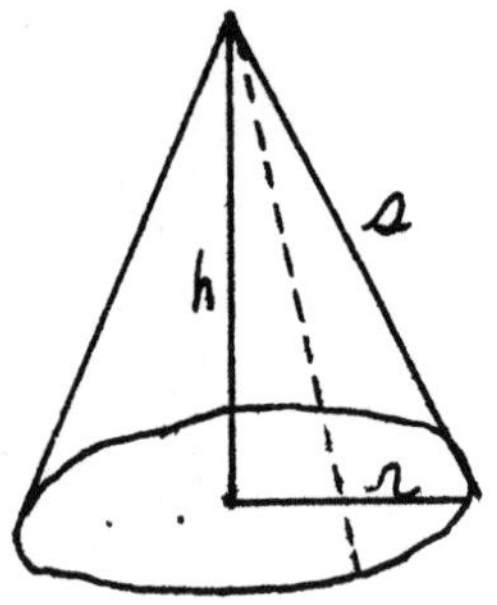

If one cuts a cone along the dotted line you have

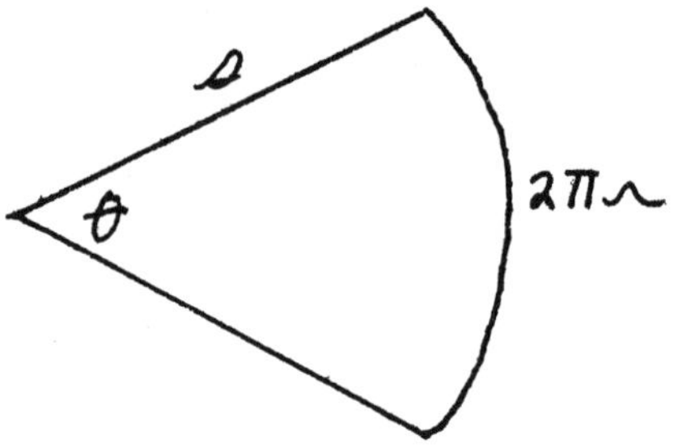

The arclength of a circular sector is $r\theta$. Thus $s\theta = 2\pi r \Rightarrow \theta = \frac{2\pi r}{s}$.

The area of a circular sector is $\frac{1}{2}s^2\theta \Rightarrow$ the area is $\frac{1}{2}s^2\frac{2\pi r}{s} = s\pi r$.

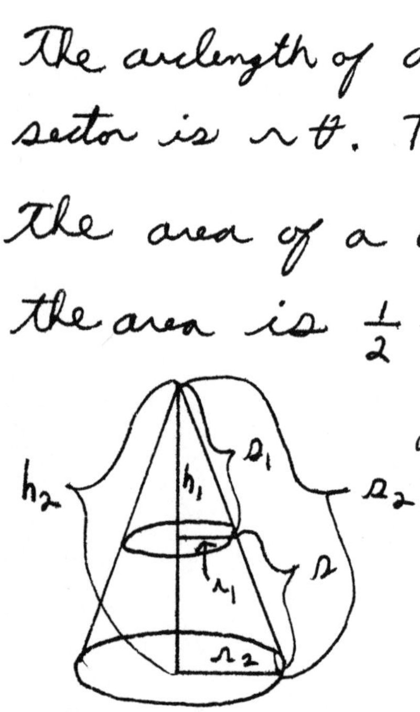

Note that $\frac{r_2}{s_2} = \frac{r_1}{s_1} \Rightarrow -r_1 s_2 + r_2 s_1 = 0$

The surface area of the frustum is

$$\pi r_1 s_1 - \pi r_2 s_2 = \pi\left[r_1 s_1 - r_1 s_2 + r_2 s_1 - r_2 s_2\right]$$

$$= \pi(r_1 + r_2)(s_2 - s_1) = \pi(r_1 + r_2)s.$$

3-9

Find the area of the surface generated by revolving the portion of the curve $y = \cosh x$, over the interval $0 \leqslant x \leqslant \ln 2$, about the x-axis.

****

Recall that surface area is given by:

$$S = \int_a^b 2\pi f(x)\sqrt{1 + [f'(x)]^2}\,dx$$

For $f(x) = \cosh x$, $f'(x) = \sinh x$ and $[f'(x)]^2 = \sinh^2 x$

So,
$$S = \int_0^{\ln 2} 2\pi \cosh x \sqrt{1 + \sinh^2 x}\,dx$$

$$= \int_0^{\ln 2} 2\pi \cosh x \sqrt{\cosh^2 x}\,dx$$

$$= \int_0^{\ln 2} 2\pi \cosh^2 x\,dx$$

$$= 2\pi \int_0^{\ln 2} \frac{1}{2}(\cosh 2x + 1)\,dx$$

$$= \pi\left(\frac{\sinh 2x}{2} + x\right)\Big]_0^{\ln 2}$$

$$= \pi\left(\frac{\sinh 2\ln 2}{2} + \ln 2\right)$$

Note that $2\ln 2 = \ln 4$ and that

$$\sinh \ln 4 = \frac{e^{\ln 4} - e^{-\ln 4}}{2} = \frac{4 - 1/4}{2} = \frac{15}{8}$$

So we have
$$S = \pi\left(\frac{15}{16} + \ln 2\right)$$

3-10

Sketch the paraboloid generated by revolving the top half of a parabola given by $y^2 = 16$ px, when $0 \leq x \leq h$, about the x-axis. Find the area of this revolution.

****

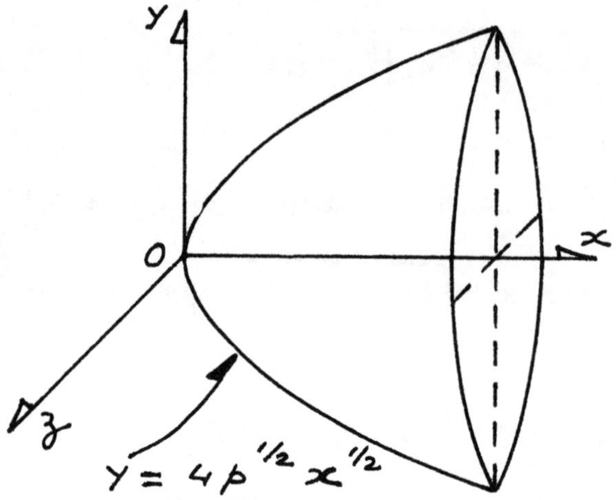

The paraboloid obtained by revolving $y^2 = 16px$ about the x-axis is shown alongside.

From the given equation, we solve for y ($y \geq 0$) as:

$$y^2 = 16px$$

$$y = \sqrt{16px}$$

$$= 4p^{1/2}x^{1/2}$$

OR $f(x) = 4p^{1/2}x^{1/2}$

The area S of the surface is given by

$$f'(x) = 4p^{1/2} \cdot \frac{1}{2} x^{-1/2}$$

$$= 2\left(\frac{p}{x}\right)^{1/2}$$

Hence, $S = 2\pi \int_a^b f(x)\sqrt{[f'(x)]^2 + 1}\, dx$

$$= 2\pi \int_0^h 4p^{1/2}x^{1/2}\sqrt{\left[2\left(\frac{p}{x}\right)^{1/2}\right]^2 + 1}\, dx$$

$$= 8\pi p^{1/2} \int_0^h x^{1/2}\sqrt{4\frac{p}{x} + 1}\, dx$$

$$= 8\pi p^{1/2} \int_0^h x^{1/2} \frac{\sqrt{4p + x}}{x^{1/2}}\, dx$$

$$= 8\pi p^{1/2} \int_0^h (4p + x)^{1/2}\, dx$$

$$= 8\pi p^{1/2} \cdot \frac{2}{3}(4p + x)^{3/2} \Big|_0^h$$

$$= \frac{16}{3}\pi p^{1/2}\left[(4p + h)^{3/2} - (4p + 0)^{3/2}\right]$$

$$= \frac{16}{3}\pi\left[\sqrt{p(4p+h)^3} - (4^{3/2}\cdot p^{3/2}\cdot p^{1/2})\right]$$

$$= \frac{16}{3}\pi\left[\sqrt{p(4p+h)^3} - 8p^2\right] \text{ sq. units}$$

Ans.

3-11

A standard formula for a sphere of radius r is

$$\text{Surface area} = 4\pi r^2.$$

Regarding the sphere as a solid of revolution, prove this formula.

****

Rotate the circle of radius r with center at the origin to get the sphere. An equation of the circle is

$$x^2 + y^2 = r^2.$$

Then

$$2x + 2y\frac{dy}{dx} = 0, \quad \frac{dy}{dx} = -\frac{x}{y},$$

$$1+\left(\frac{dy}{dx}\right)^2 = 1 + \frac{x^2}{y^2} = \frac{y^2+x^2}{y^2} = \frac{r^2}{y^2}.$$

Therefore

$$\text{Surface area} = \int_a^b 2\pi y\sqrt{1+\left(\frac{dy}{dx}\right)^2}\,dx$$

$$= 2\pi\int_{-r}^{r} y\sqrt{\frac{r^2}{y^2}}\,dx = 2\pi\int_{-r}^{r} y\left(\frac{r}{y}\right)dx$$

$$= 2\pi\int_{-r}^{r} r\,dx = 2\pi r x\Big|_{-r}^{r} = 2\pi r\,(r-(-r)) = 4\pi r^2$$

3-12

Find the surface area generated when the curve y = cosh x [0≤ x ≤ 1] is rotated around the y-axis.

****

$$S.A. = \int_{x_1}^{x_2} 2\pi x \sqrt{1+y'^2}\, dx$$

$$= \int_0^1 2\pi x \sqrt{1+\sinh^2 x}\, dx$$

$$= 2\pi \int_0^1 x \cosh x\, dx. \text{ Integrate by Parts}$$

$$u = x \qquad dv = \cosh x\, dx$$
$$du = dx \qquad v = \sinh x$$

$$= 2\pi \left[x \sinh x \Big|_0^1 - \int_0^1 \sinh x\, dx \right]$$

$$= 2\pi(\sinh 1 - \cosh 1 + \cosh 0)$$

$$= 2\pi\left(\frac{e^1 - e^{-1}}{2} - \frac{e^1 + e^{-1}}{2} + 1 \right)$$

$$= 2\pi\left(1 - \frac{1}{e}\right)$$

VOLUMES BY SLICING

3-13

The base of a solid is the region in the xy-plane enclosed by the curves $y = e^x$, $y = 1$, and $x = 2$. Every cross section of the solid taken perpendicular to the x-axis is a square. Find the volume of the solid.

****

$$\text{Vol.} = \int_0^2 A(x)\,dx$$

where $A(x)$ is the area of the cross section at x.

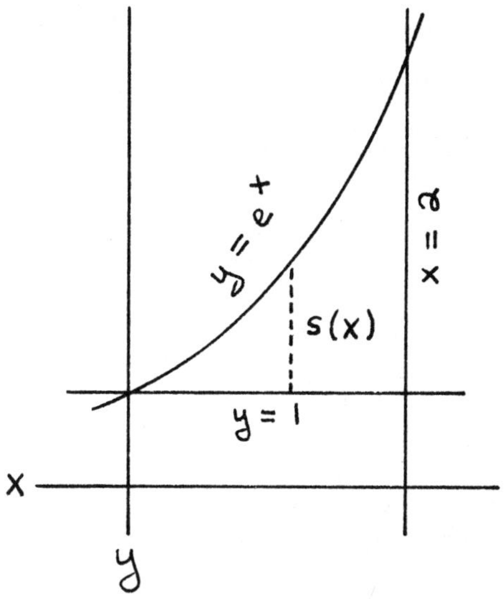

From the diagram, the length of the side of the square cross section at x is given by

$$s(x) = e^x - 1.$$

So, $A(x) = (s(x))^2 = (e^x - 1)^2 = e^{2x} - 2e^x + 1$.

$$\therefore \text{Vol.} = \int_0^2 (e^{2x} - 2e^x + 1)\,dx = \frac{1}{2}e^{2x} - 2e^x + x\Big|_0^2$$

$$= \left(\frac{1}{2}e^4 - 2e^2 + 2\right) - \left(\frac{1}{2} - 2\right) = \frac{1}{2}e^4 - 2e^2 + \frac{7}{2}.$$

3-14

The base of a solid is the region in the xy-plane bounded by the curves $y = x^2/3$ and $y = 3$. Cross-sections of the solid perpendicular to the y-axis are squares rising vertically out of the xy-plane with their bottom edges in the xy-plane. What is the volume of this solid?

****

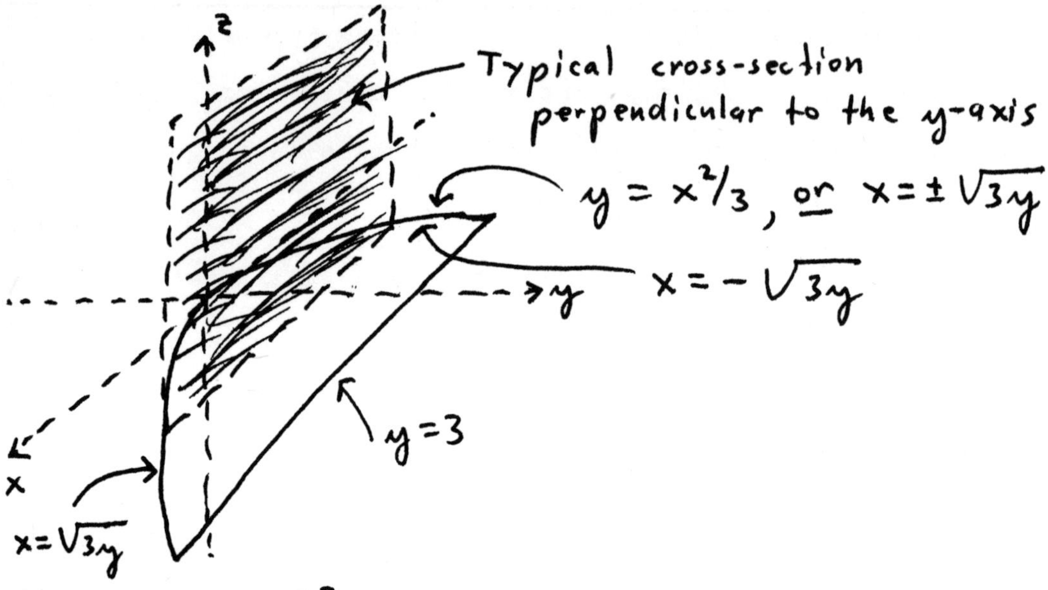

$$\text{Volume} = \int_0^3 \left[\sqrt{3y} - (-\sqrt{3y})\right]^2 dy$$

$$= \int_0^3 (2\sqrt{3y})^2\, dy = \int_0^3 12y\, dy = 6y^2\Big|_0^3$$

$= 54$ cubic units

3-15

Which has the larger volume the ellipsoid formed by revolving $\frac{x^2}{a^2} + \frac{y^2}{b^2} = 1$ about the x-axis or the ellipsoid formed by revolving $\frac{x^2}{a^2} + \frac{y^2}{b^2} = 1$ about the y-axis. $(a > b)$

****

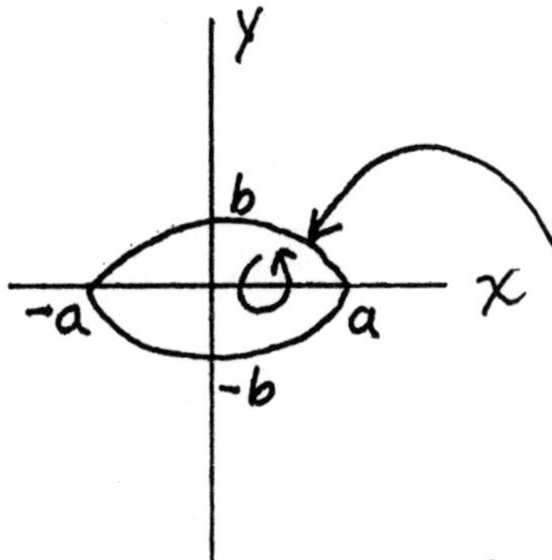

$\frac{x^2}{a^2} + \frac{y^2}{b^2} = 1$ an ellipse

$b^2x^2 + a^2y^2 = a^2b^2$

$y = \frac{b\sqrt{a^2 - x^2}}{a}$

We will use the slicing method

$$V = 2\pi\int_0^a \left(\frac{b\sqrt{a^2-x^2}}{a}\right)^2 dx = \frac{2b^2\pi}{a^2}\int_0^a a^2 - x^2\, dx = \frac{2b^2\pi}{a^2}\left\{a^2x - \frac{x^3}{3}\right|_0^a =$$

$$\frac{2b^2\pi}{a^2}\left\{\frac{2a^3}{3}\right\} = \frac{4ab^2\pi}{3}.$$

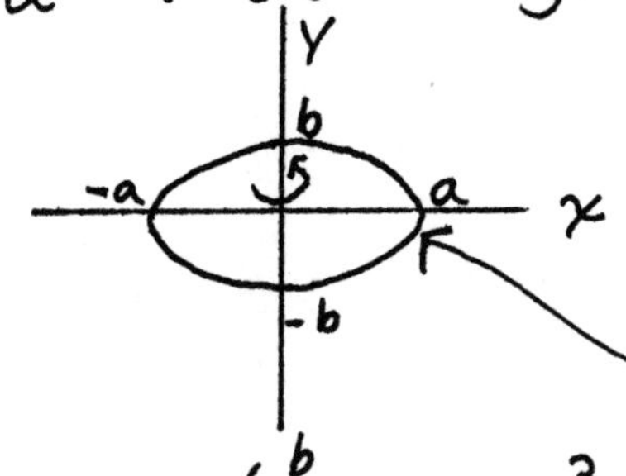

$\frac{x^2}{a^2} + \frac{y^2}{b^2} = 1$

$b^2x^2 + a^2y^2 = a^2b^2$

$x = \frac{a\sqrt{b^2 - y^2}}{b}$

$$V = 2\pi\int_0^b \left(\frac{a\sqrt{b^2-y^2}}{b}\right)^2 dy = \frac{2a^2\pi}{b^2}\int_0^b (b^2 - y^2)\, dy = \frac{2a^2\pi}{b^2}\left\{b^2y - \frac{y^3}{3}\right|_0^b =$$

$$\frac{2a^2\pi}{b^2}\left\{\frac{2b^3}{3}\right\} = \frac{4a^2b\pi}{3}.$$ Now $a > b \Rightarrow \frac{4a^2b\pi}{3} > \frac{4ab^2\pi}{3}$.

The largest volume is formed by revolving the ellipse about the y-axis.

3-16

Find the volume of the solid generated when the region in the first quadrant bounded by the curve $y = 4x - x^3$ and the x-axis is rotated about the x-axis.

****

FIRST, SKETCH THE GRAPH.

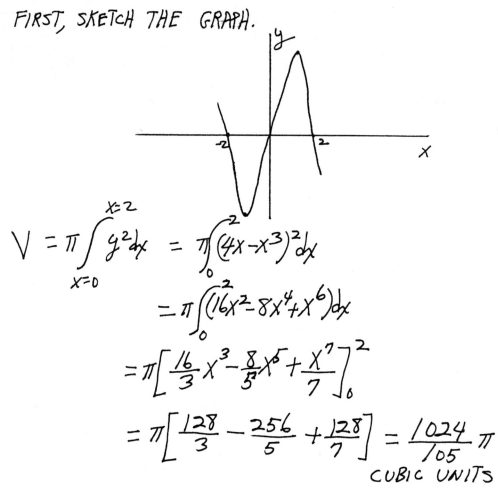

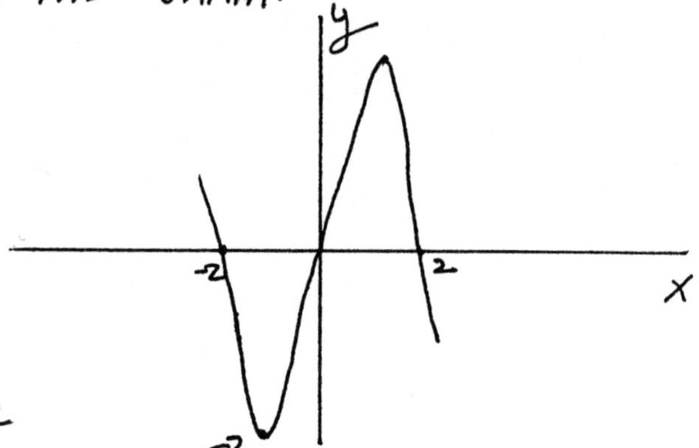

$$V = \pi \int_{x=0}^{x=2} y^2\,dx = \pi \int_0^2 (4x - x^3)^2\,dx$$

$$= \pi \int_0^2 (16x^2 - 8x^4 + x^6)\,dx$$

$$= \pi \left[\frac{16}{3}x^3 - \frac{8}{5}x^5 + \frac{x^7}{7} \right]_0^2$$

$$= \pi \left[\frac{128}{3} - \frac{256}{5} + \frac{128}{7} \right] = \frac{1024}{105}\pi$$

CUBIC UNITS

3-17

Find the volume generated by revolving the region bounded by the x axis, $x = 0$, $x = 1$, and $y = e^x$ about the x axis.

****

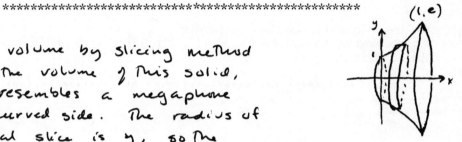

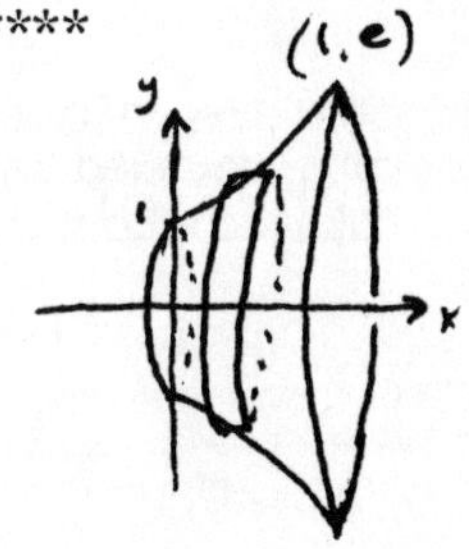

Use the volume by slicing method to find the volume of this solid, which resembles a megaphone with curved side. The radius of a typical slice is y, so the volume is

$$\text{Volume} = \int_a^b \pi y^2\,dx = \pi\int_0^1 (e^x)^2\,dx = \pi\int_0^1 e^{2x}\,dx$$

$$= \frac{\pi}{2}\int_0^1 e^{2x}(2)\,dx = \frac{\pi}{2}e^{2x}\Big|_0^1 = \frac{\pi}{2}(e^2 - e^0)$$

$$= \frac{\pi}{2}(e^2 - 1)$$

3-18

Find the volume of the solid whose base is the interior of the ellipse $x^2 + 4y^2 = 4$, and whose cross sections perpendicular to the x-axis are semi-circles.

****

$$\text{VOLUME} = \int_{-2}^{2} \frac{1}{2}\pi\left(\frac{4 - x^2}{4}\right)dx$$

$$= \frac{\pi}{8}\int_{-2}^{2}(4 - x^2)\,dx$$

$$= \frac{\pi}{8}\left[4x - \frac{x^3}{3}\right]_{-2}^{2}$$

$$= \frac{\pi}{8}\left(8 - \frac{8}{3} + 8 - \frac{8}{3}\right) = \frac{4\pi}{3}$$

VOLUMES BY SHELLS

3-19

Find the volume of the solid generated when the region in the first quadrant bounded by the curve $x = y^2 + 2$, the line $x = 3$ and the x-axis is rotated about the line $x = 4$.

****

FIRST, SKETCH THE GRAPH.

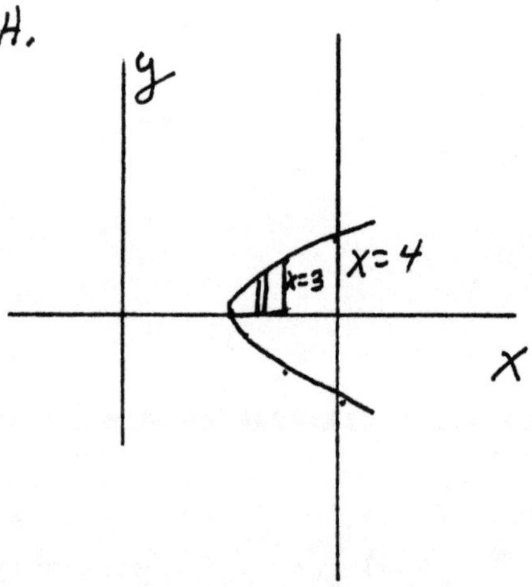

$$V = \int_{x=2}^{x=3} 2\pi(4-x)y\,dx$$

$$= \int_2^3 2\pi(4-x)\sqrt{x-2}\,dx$$

LET $u = \sqrt{x-2}$

$u^2 = x-2$

$2u\,du = dx$

$$= 2\pi\int_0^1 (2-u^2)u(2u\,du) = 4\pi\int_0^1 (2u^2-u^4)\,du$$

$$= 4\pi\left[\frac{2}{3}u^3 - \frac{u^5}{5}\right]_0^1 = 4\pi\left[\frac{2}{3}-\frac{1}{5}\right] = \frac{28}{15}\pi$$

CUBIC UNITS

4-33

A square is to be cut from each corner of a piece of paper which is 8 cm. by 10 cm. and the sides are to be folded up to create an open box. What should the side of the square be for maximum volume?

****

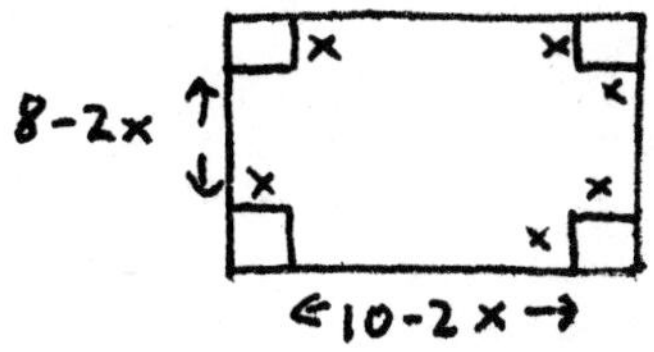

$V_{box} = l \cdot w \cdot h$

$V_{box} = (10-2x)(8-2x)x = 80x - 36x^2 + 4x^3$

Maximum volume will occur where $\frac{dV}{dx} = 0$

$$\frac{dV}{dx} = 80 - 72x + 12x^2 = 0$$

$$x = 3 + \frac{\sqrt{21}}{3} \text{ or } 3 - \frac{\sqrt{21}}{3}$$

To determine which of these will give maximum volume, use 2nd derivative test.

$$\frac{d^2V}{dx^2} = -72 + 24x$$

If $x = 3 + \frac{\sqrt{21}}{3}$ this is $-72 + 24\left(3 + \frac{\sqrt{21}}{3}\right) = 8\sqrt{21}$ POS. MIN.

If $x = 3 - \frac{\sqrt{21}}{3}$ this is $-72 + 24\left(3 - \frac{\sqrt{21}}{3}\right) = -8\sqrt{21}$ NEG. MAX

Answer is $3 - \frac{\sqrt{21}}{3} \approx 1.47$ cm

4-34

A carpenter wishes to build a bin with lid in the corner of a room, utilizing the corner walls and floor. What dimensions will hold 8 cubic feet using the least plywood, if the base is square?

****

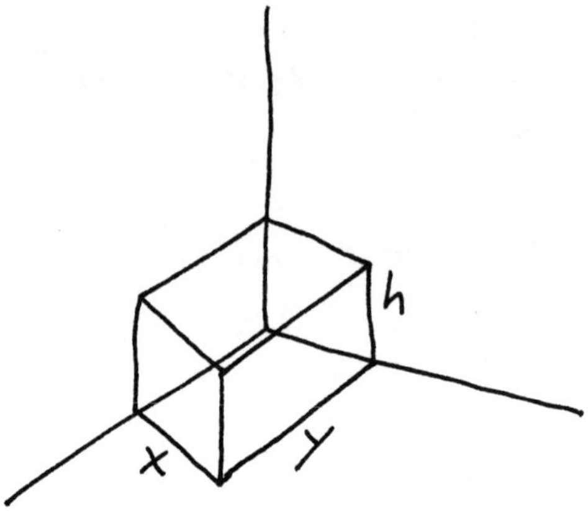

The plywood required is $P = xh + yh + xy$ $= 2xh + x^2$ since the base is square. Since $V = 8 = xyh$ and $x = y$, $x^2h = 8$. Thus $h = 8/x^2$ and $P = x^2 + 16/x$.

$P' = 2x - 16/x^2 = 0$ when $2x^3 = 16$, $x = 2$.

Thus the most economical bin is a $2 \times 2 \times 2$ cube.

4-35

Consider all rectangles of the type shown in the figure for $0 < t < 1$. What value(s) of t (if any) will yield the rectangles of largest and smallest area?

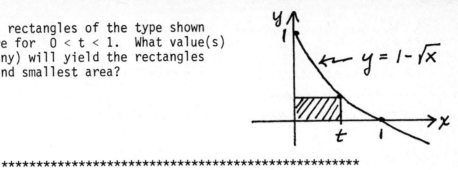

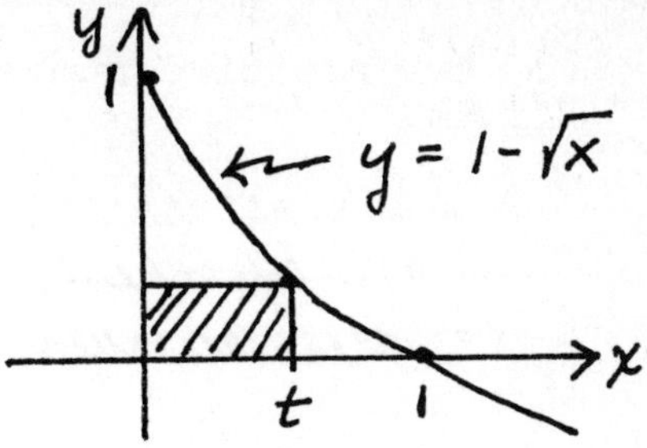

****

The area A of a typical rectangle is given by:

$$A = t(1-\sqrt{t}) = t - t^{3/2}$$

To find the extreme values of A, we set the derivative equal to 0:

$$\frac{dA}{dt} = 1 - \frac{3}{2}t^{1/2} = 0 \quad \text{when} \quad \frac{3}{2}t^{1/2} = 1$$

$$\text{or} \quad t^{1/2} = \frac{2}{3} \quad \text{or} \quad t = 4/9$$

So, an extreme value of A occurs when $t = \frac{4}{9}$ and it is intuitively clear that this gives a maximum of A. To check this, use the second derivative test:

$$\frac{d^2A}{dt^2} = -\frac{3}{4}t^{-1/2} = -\frac{3}{4\sqrt{t}}. \quad \text{Hence } \frac{d^2A}{dt^2} < 0$$

for all $t > 0$, which confirms that:

$t = 4/9$ gives a maximum of A

Since the degenerate rectangles with area 0 that occur for $t=0$ and $t=1$ are excluded by hypothesis $(0<t<1)$:

there is no rectangle of minimum area.

4-36

Find two positive numbers whose sum is 6 whose product is a maximum.

****

x = one number

$6-x$ = other number. $P = x(6-x) = 6x - x^2$

extreme values are located where 1st derivative = 0

$$P' = 6 - 2x$$

$$P' = 0 \Rightarrow 6 - 2x = 0 \Rightarrow x = 3$$

Use 2nd derivative test to determine maximum or minimum.

$$P'' = -2$$

$P''(3) = -2$ negative value $\Rightarrow$ $x = 3$ represents a maximum

so two numbers are $x=3$ and $6-x=3$.

THE FIRST DERIVATIVE TEST

4-37

Find all relative maximums and minimums of $f(x) = 2x^3 - 3x^2 - 12x + 5$ using the first derivative test.

****

Possible relative extremums exist where $f'(x)=0$.
$f'(x) = 6x^2 - 6x - 12 = 6(x^2 - x - 2) = 6(x-2)(x+1)$, so $f'(x)=0$ when $x=2$ or $x=-1$.

Choosing a value less than $x=-1$, say $x=-2$, we see that $f'(-2) = 24+12-12 = 24 > 0$, which means $f(x)$ is increasing for $x<-1$. Choosing a value between $x=-1$ and $x=2$, say $x=0$, we see that $f'(0) = 0-0-12 = -12 < 0$, which means that $f(x)$ is decreasing for $-1<x<2$. Choosing a value of x greater than $x=2$, say $x=3$, we see that $f'(3) = 54-18-12 = 24 > 0$, which means $f(x)$ is increasing for $x>2$.

Conclusion: Since $f(x)$ changes from increasing to decreasing at $x=-1$, a relative maximum exists at $x=-1$, and is the point $(-1, 12)$.

Since $f(x)$ changes from decreasing to increasing at $x=2$, a relative minimum exists at $x=2$, and is the point $(2, -15)$.

4-38

Let $f(x)=\frac{1}{3}x^3 - x$. Use the first derivative test to determine the following: (a) Where f(x) is increasing and (b) where f(x) is decreasing.

****

(a) $f'(x) = x^2-1$. Since f is increasing when $f'(x)>0$, let $f'(x)>0$. Then $x^2-1>0$ and $(x+1)(x-1)>0$. Hence either

$x+1>0$ and $x-1>0$
$x>-1$ and $x>1$
from which we get $x>1$

OR

$x+1<0$ and $x-1<0$
$x<-1$ and $x<1$
from which we get $x<-1$

Hence f is increasing when $x>1$ when $x<-1$.

(b) Since f is decreasing when $f'(x)<0$, let $f'(x)<0$ and solve the resulting inequality.
Then $x^2-1<0$ so that $(x+1)(x-1)<0$.
Hence either

$x+1<0$ and $(x-1)>0$
$x<-1$ and $x>1$
from which we get No solution

OR

$x+1>0$ and $x-1<0$
$x>-1$ and $x<1$
from which we get $-1<x<1$.

So f is decreasing if $-1<x<1$.

Therefore f is increasing when $x>1$ and when $x<-1$ but decreasing when $-1<x<1$.

■■ **4-39**

Find and describe all local extrema of $f(x) = x^{5/3} - 5x^{2/3}$

****

$$f(x) = x^{5/3} - 5x^{2/3}$$

$$f'(x) = \tfrac{5}{3}x^{2/3} - \tfrac{10}{3}x^{-1/3} = \tfrac{5}{3}\left[x^{2/3} - \frac{2}{x^{1/3}}\right]$$

$$= \tfrac{5}{3}\left[\frac{x-2}{x^{1/3}}\right]$$

so $f'(x) = \tfrac{5}{3}\left(\frac{x-2}{x^{1/3}}\right)$

now we have critical points when $f'(x) = 0$ or when $f'(x)$ is undefined

firstly $f'(x) = 0$ when $\frac{x-2}{x^{1/3}} = 0 \rightarrow x - 2 = 0$ when $x = 2$

secondly $f'(x)$ is undefined when $x^{1/3} = 0$, when $x = 0$

$\tfrac{5}{3}(x-2)$ - - - - - - - - - - - - + + + + ++

$x^{1/3}$ - - - - - + + + + + + + + + ++ + +

$f'(x)$ +   0   -   2   +

notice $f(0) = 0$ so at $(0,0)$ we have a vertical tangent, or 'elbow'

f(x) has a relative minimum at x=2, and a relative maximum at x=0

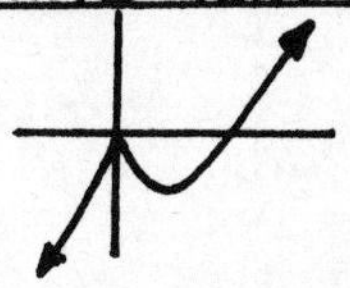

THE SECOND DERIVATIVE TEST

4-40

Let $f(x)=\frac{1}{3}x^3 - x$. Find the critical values of f and then use the second derivative to determine whether these critical values are relative max. or relative minimum.

****

$f'(x) = x^2 - 1$. So let $f'(x)=0$ to get $x = \pm 1$ for critical values.
$f''(x) = 2x$ and since $f''(1) = 2 > 0$, the point $(1, -\frac{2}{3})$ is a relative minimum point. Since $f''(-1) = -2 < 0$, the point $(-1, \frac{2}{3})$ is a relative maximum point.

4-41

Find all relative maximums and minimums of $f(x) = x^4 - (4/3)x^3 - 12x^2 + 1$ using the second derivative test.

****

Possible relative extremums exist where $f'(x) = 0$.

$f'(x) = 4x^3 - 4x^2 - 24x = 4x(x^2 - x - 6) = 4x(x-3)(x+2)$,

so $f'(x) = 0$ when $x = 0$, $x = 3$, or $x = -2$.

$f''(x) = 12x^2 - 8x - 24$

$f''(0) = 0 - 0 - 24 = -24 < 0$, so a relative maximum exists at $x = 0$, and is the point $(0, 1)$.

$f''(3) = 108 - 24 - 24 = 60 > 0$, so a relative minimum exists at $x = 3$, and is the point $(3, -62)$.

$f''(-2) = 48 + 16 - 24 = 40 > 0$, so a relative minimum exists at $x = -2$, and is the point $(-2, -61/3)$.

RELATED RATES

4-42

A cylindrical can is undergoing a transformation in which the radius and height are varying continuously with time t. The radius is increasing at 4 in/min, while the height is decreasing at 10 in/min. Is the volume of the can increasing or decreasing, and at what rate, when the radius is 3 inches and the height is 5 inches?

****

Volume of can = $V = \pi r^2 h$

$$\frac{dV}{dt} = \pi\left[2rh\, dr/dt + r^2\, dh/dt\right]$$

$$= \pi\left[8rh - 10r^2\right], \text{ since } dr/dt = +4 \text{ in/min}$$

$$\text{and } dh/dt = -10 \text{ in/min}$$

$$\therefore \left.\frac{dV}{dt}\right|_{\substack{r=3\\h=5}} = \pi\left[8(3)(5) - 10(3^2)\right]$$

$$= \pi[120 - 90] = 30\pi$$

The volume of the can is <u>increasing</u> at the rate of 30π cubic inches per minute, when the radius is 3 inches and the height is 5 inches.

4-43

A frugal young man has decided to extract one of his teeth by tying a stout rubber band from his tooth to the chain on a garage door opener which runs on a horizontal track 3 feet above his mouth. If the garage door opener moves the chain at 1/4 ft/sec, how fast is the rubber band expanding when it is stretched to a length of 5 feet?

****

LET THE VARIABLES BE CHOSEN AS SHOWN IN ELABORATE FIGURE.

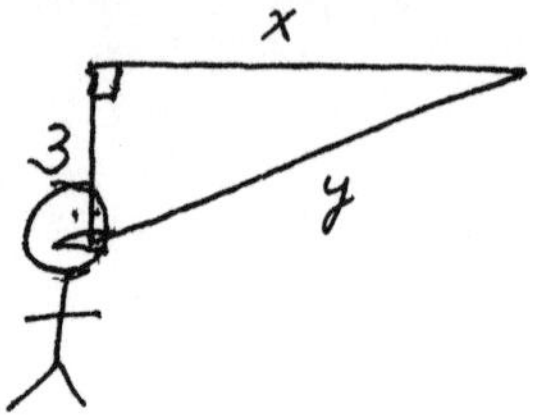

WE KNOW THAT $y^2 = x^2 + 9$ AND THAT $\frac{dx}{dt} = 1/4$. WE NEED $\frac{dy}{dt}$ WHEN $y = 5$.

DIFFERENTIATING WITH RESPECT TO t, WE HAVE

$$2y\frac{dy}{dt} = 2x\frac{dx}{dt}$$

$$y\frac{dy}{dt} = x\frac{dx}{dt}.$$

WHEN $y = 5$, WE HAVE

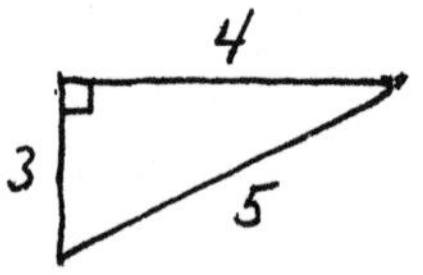

SO $5\frac{dy}{dt} = 4 \cdot \frac{dx}{dt} = 4 \cdot \frac{1}{4} = 1$

$$\frac{dy}{dt} = \frac{1}{5}.$$

THE RUBBER BAND IS EXPANDING AT $\frac{1}{5}$ FT/SEC.

4-44

Two straight roads intersect at right angles in Newtonville. Car A is on one road moving toward the intersection at a speed of 50 m.p.h. Car B is on the other road moving away from the intersection at a speed of 30 m.p.h. When car A is 2 miles from the intersection and car B is 4 miles from the intersection:

(a) How fast is the distance between the cars changing?

(b) Are the cars getting closer together or farther apart?

****

(a)

$$D^2 = x^2 + y^2$$

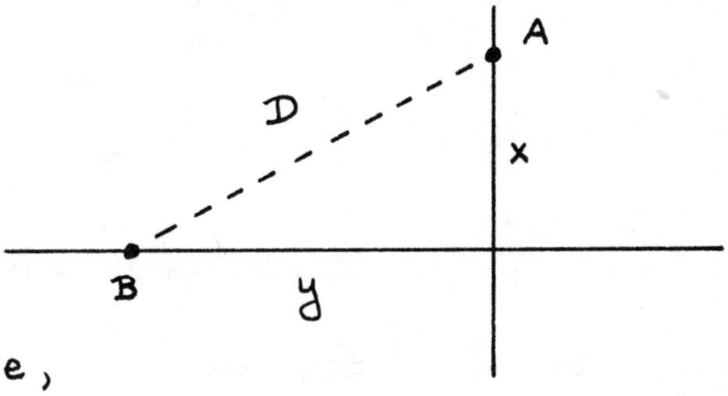

Differentiating with respect to time,

$$2DD' = 2xx' + 2yy', \quad \text{so} \quad D' = \frac{xx' + yy'}{D}$$

At the time in question,

$$x = 2, \quad y = 4, \quad D = \sqrt{x^2 + y^2} = \sqrt{20}, \quad \text{and}$$

$$x' = -50 \ (x \text{ decreasing}), \quad y' = +30 \ (y \text{ increasing})$$

$$\therefore \quad D' = \frac{(2)(-50) + (4)(30)}{\sqrt{20}} = \sqrt{20} \text{ m.p.h.}$$

(b) $D' > 0$ so D is increasing and the cars are getting farther apart.

4-45 ■■■■ ■■ ■■■ ■■■ ■■■ ■■■ ■■■ ■■■ ■■■ ■■■ ■■■ ■■■ ■■■ ■■■

The length of a rectangle is increasing at the rate of 7 ft/sec, while the width is decreasing at the rate of 3 ft/sec. At one time, the length is 12 feet and the diagonal is 13 feet. At this time find the rate of change in
a) the area
b) the perimeter
and tell whether each is increasing or decreasing.

****

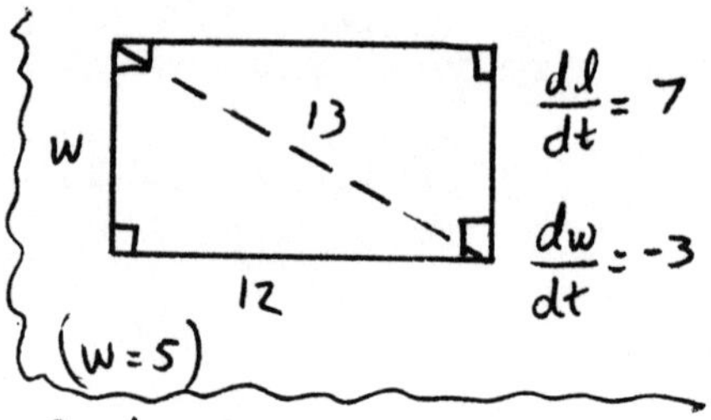

a) $A = l \cdot w$

$\dfrac{dA}{dt} = w \cdot \dfrac{dl}{dt} + l \cdot \dfrac{dw}{dt}$

$\dfrac{dl}{dt} = 7$

$\dfrac{dw}{dt} = -3$

$(w=5)$

when $l = 12$, $w = 5$, the rate of change of the area

is $\dfrac{dA}{dt} = (7)(5) + (12)(-3) = \boxed{-1 \text{ sq. ft/sec}}$

and it is $\boxed{\text{decreasing}}$

b) $P = 2l + 2w$

$\dfrac{dP}{dt} = 2 \cdot \dfrac{dl}{dt} + 2 \dfrac{dw}{dt}$

$\dfrac{dP}{dt} = (2)(7) + (2)(-3) = \boxed{8 \text{ ft/sec}}$

and it is $\boxed{\text{increasing}}$

4-46

A particle starts at the origin and moves along the parabola $y = x^2$ such that it distance from the origin increases at 4 units per second. How fast is its x-coordinate changing as it passes through the point (1,1) ?

****

Let s = distance of particle from (0,0)

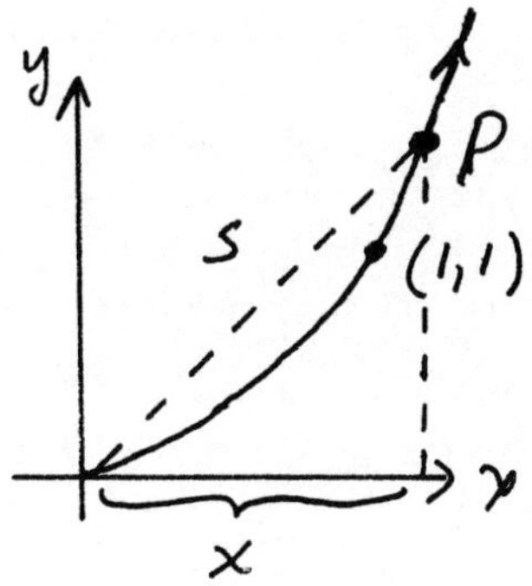

Given: $\frac{ds}{dt} = 4$

To find: $\left.\frac{dx}{dt}\right]_{x=1} = ?$

We need a relation between s and x

In general, $s^2 = x^2 + y^2$

Since the particle is restricted to $y = x^2$:

$$s^2 = x^2 + (x^2)^2 = x^2 + x^4$$

Differentiating both sides of the above equation with respect to t:

$$2s \cdot \frac{ds}{dt} = 2x \cdot \frac{dx}{dt} + 4x^3 \frac{dx}{dt}$$

or: $s \frac{ds}{dt} = x \cdot \frac{dx}{dt} + 2x^3 \cdot \frac{dx}{dt}$.

By hypothesis $ds/dt = 4$. Also, when $x = 1$, $s = \sqrt{2}$

Hence: $\sqrt{2} \cdot 4 = 1 \cdot \left.\frac{dx}{dt}\right]_{x=1} + 2 \cdot 1^3 \cdot \left.\frac{dx}{dt}\right]_{x=1}$

So: $\left.\frac{dx}{dt}\right]_{x=1} = \frac{4}{3}\sqrt{2}$ units/sec.

4-47

When a stone is dropped in a pool, a circular wave moves out from the point of impact at a rate of six inches per second. How fast is the area enclosed by the wave increasing when the wave is two inches in radius? (Recall: $A = \pi r^2$, where A = area, r = radius).

****

Given $A = \pi r^2$, we get $dA = \pi \cdot 2r \cdot dr$.

In the problem, $dr = 6$ and $r = 2$.

Thus $dA = \pi \cdot 2 \cdot 2 \cdot 6 = 24\pi$ sq. in./sec.

4-48

Suppose that $y = 2x^2 - 3x + 1$.

(a) Find and simplify a formula for the y increment, Δy.

(b) Find a formula for the y differential, dy.

****

(a) $\Delta y = 2(x+\Delta x)^2 - 3(x+\Delta x) + 1 - [2x^2 - 3x + 1]$

$= 2x^2 + 4x(\Delta x) + 2(\Delta x)^2 - 3x - 3(\Delta x) + 1 - 2x^2 + 3x - 1$

$= 4x(\Delta x) + 2(\Delta x)^2 - 3(\Delta x)$

(b) $dy = y'\,dx = (4x - 3)\,dx$

4-49

Using differentials, find an approximation for the cube root of 30.

****

Let $f(x) = \sqrt[3]{x} = x^{1/3}$. If Δx is small and nonzero, then $f(x+\Delta x) \approx f(x) + f'(x)\Delta x$.

$\sqrt[3]{30} = f(30) = f(27+3)$. Therefore $x = 27$ and $\Delta x = 3$. Since $f'(x) = \frac{1}{3}x^{-2/3} = \frac{1}{3x^{2/3}}$,

$$\sqrt[3]{30} = f(27+3) \approx f(27) + f'(27)(3)$$

$$\approx 3 + \frac{1}{3(9)} \cdot 3$$

$$\approx 3 + \frac{1}{9} = 3.\overline{1}$$

A calculator check shows $\sqrt[3]{30} \approx 3.1072$.

4-50

A particle moves along a path described by $y = x^2$. At what point along the curve are x and y changing at the same rate? Find this rate if at that time t we have $x = \sin(t)$ and $y = \sin^2 t$.

****

$\frac{dy}{dt} = 2x\frac{dx}{dt}$ $\quad \frac{dy}{dt} = \frac{dx}{dt} \Rightarrow \frac{dx}{dt} = 2x\frac{dx}{dt} \Rightarrow 1 = 2x$

$\Rightarrow \frac{1}{2} = x$

at $x = 1/2$, $y = (1/2)^2 = 1/4$ so point is $(1/2, 1/4)$.

now for $x = \sin t$, $\frac{dx}{dt} = \cos t$

at $x = 1/2$, $\sin t = 1/2 \Rightarrow t = \pi/6$ so $\frac{dx}{dt} = \cos \pi/6 = \sqrt{3}/2$

4-51

Two of the cutest little puppies you've ever seen begin running from the same point. One of them--a big-eyed little rascal with the endearing habit of cocking his head to one side--romps due east at 5 miles per hour. Frisky, though pudgy, the other darling little tyke gambols directly northward at only 4 miles per hour. At what rate is the distance between the puppies changing after 2 hours?

****

LET THE VARIABLES BE CHOSEN AS SHOWN IN THE FIGURE.

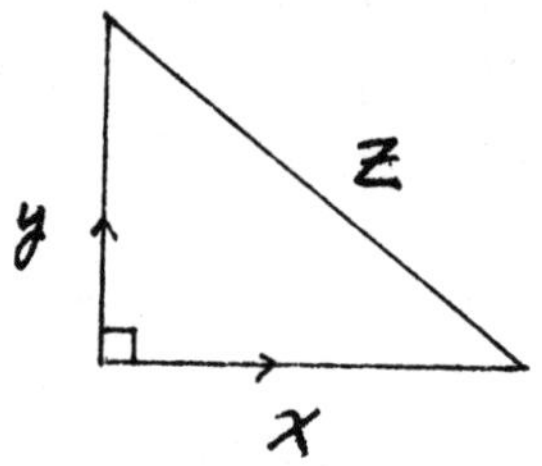

WE KNOW THAT

$$x^2 + y^2 = z^2$$

AND THAT

$$\frac{dx}{dt} = 5 \qquad \frac{dy}{dt} = 4.$$

WE WANT $\frac{dz}{dt}$ AT $t = 2$.

DIFFERENTIATING IMPLICITLY WITH RESPECT TO t,

$$2x\frac{dx}{dt} + 2y\frac{dy}{dt} = 2z\frac{dz}{dt}$$

$$x\frac{dx}{dt} + y\frac{dy}{dt} = z\frac{dz}{dt}.$$

AT $t = 2$, WE HAVE

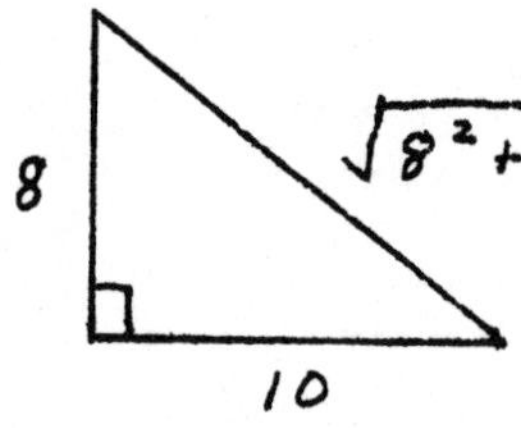

$$\sqrt{8^2 + 10^2} = \sqrt{164} = 2\sqrt{41}$$

SO

$$10 \cdot 5 + 8 \cdot 4 = 2\sqrt{41}\frac{dz}{dt}$$

$$82 = 2\sqrt{41}\frac{dz}{dt}$$

$$\frac{dz}{dt} = \frac{82}{2\sqrt{41}} = \frac{82\sqrt{41}}{2\sqrt{41}\sqrt{41}} = \sqrt{41}.$$

THE DISTANCE IS CHANGING AT $\sqrt{41}$ MI/HR.

4-52

A mothball shrinks in such a way that its radius decreases by 1/6th inch per month. How fast is the volume changing when the radius is 1/4th inch? Assume mothball is spherical.

****

"How fast" asks us to find the change in volume compared to the change in time; ie $\frac{dV}{dt}$

$$\frac{dV}{dt} = \frac{dV}{dr} \cdot \frac{dr}{dt}$$

Since $V = \frac{4}{3}\pi r^3$, $\frac{dV}{dr} = 4\pi r^2$

It's given that $\frac{dr}{dt} = \frac{-1}{6}$

So $\frac{dV}{dt} = 4\pi r^2 \cdot \frac{-1}{6} = \frac{-2\pi r^2}{3}$;

when $r = \frac{1}{4}$ this is $\frac{-\pi}{24}$ "/month

4-53

The electric resistance of a certain resistor as a function of temperature is given by $R = 6.000 + 0.002T^2$, where R is measured in Ohms and T in degrees Celsius. If the temperature is decreasing at the rate of 0.2°C per second, find the rate of change of resistance when T = 38°C.

****

GIVEN: $\frac{dT}{dt} = -0.2°C/\text{second}$

$$\left.\frac{dR}{dt}\right|_{T=38°} = \left.2(.002)T\frac{dT}{dt}\right|_{T=38°} = .004(38)(-.2)$$

$$= -0.0304 \text{ ohms/second}$$

4-54

A man 6 feet tall walks at the rate of 200 feet per minute towards a street light which is 18 feet above the ground. At what rate is the tip of his shadow moving?

****

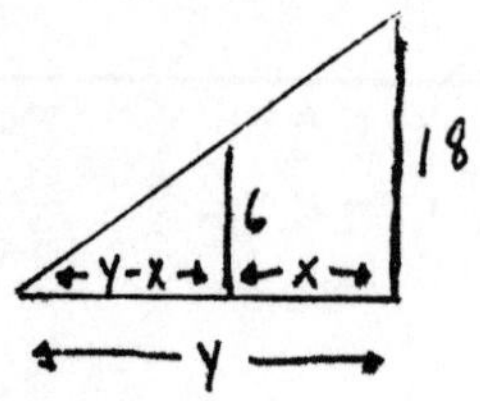

X = distance from man to light
Y = distance from tip of shadow to light.
$Y-X$ = distance from tip of shadow to man.

USING SIMILAR TRIANGLES $\frac{18}{Y} = \frac{6}{Y-X}$

Then, $18Y - 18X = 6Y$

OR $12Y = 18X$

OR $Y = \frac{3}{2}X$

DIFFERENTIATING BOTH SIDE, WITH RESPECT TO TIME YIELDS $\frac{DY}{DT} = \frac{3}{2}\frac{DX}{DT}$

THEREFORE $\frac{DY}{DT} = \frac{3}{2}(200) = 300\ \frac{FT.}{MIN.}$

4-55

A streetlight is 12 feet high. A moth is 2 feet from the lamppost and is flying straight up at 1 foot per second. How fast is its shadow moving along the ground when it's 11 feet off the ground?

****

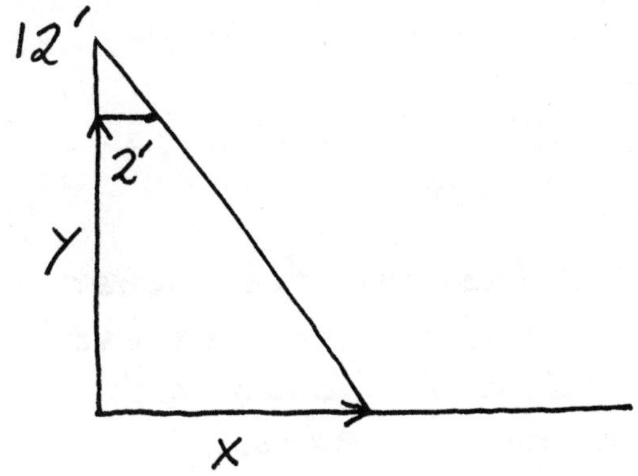

We want $\frac{dx}{dt}$ when $y = 11$. Using similar triangles, $\frac{x}{12} = \frac{2}{12-y}$, $x = \frac{24}{12-y}$.

Thus $\frac{dx}{dt} = 24 \frac{d}{dt}\left(\frac{1}{12-y}\right)$

$$= 24 \frac{(-1)(-1)}{(12-y)^2} \frac{dy}{dt}.$$

When $y = 11$, $\frac{dx}{dt} = 24$ ft/sec since $\frac{dy}{dt}$ is always 1 ft/sec.

4-56

Mr. Cooper is standing on the top of his 16 foot ladder when he realizes that the ladder is slipping down the side of the building. He decides that the base of the ladder is moving away from the bottom of the building at a rate of 2 feet per second when it is 3 feet from the bottom of the building. How fast is Mr. Cooper falling at that instant?

****

Let x be the distance of the base of the ladder from the building and y be the height of the ladder up on the building at time t. See the diagram below.

Given: $\frac{dx}{dt} = 2$ feet per second when $x = 3$.

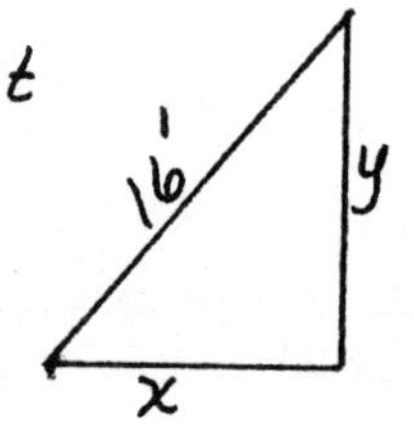

We want to find $\frac{dy}{dt}$ when $x = 3$ feet.

Since the ladder forms a rt. triangle, $x^2+y^2 = 256$. So if $x = 3$, $9+y^2=256$ and $y = \sqrt{247}$.

Differentiating $x^2+y^2=256$ with respect to t we get $x\frac{dx}{dt} + y\frac{dy}{dt} = 0$. Replace x by 3, $\frac{dx}{dt}$ by 2 and y by $\sqrt{247}$ to obtain $\frac{dy}{dt} = -\frac{6}{\sqrt{247}}$ feet per second when $x = 3$. Since the sign is negative, Mr. Cooper is <u>falling</u> at the rate of $\frac{6}{\sqrt{247}}$ feet per second.

CONCAVITY AND POINTS OF INFLECTION

4-57

Determine a so that the function $f(x) = x^2 + \frac{a}{x}$ has an inflection point at $x = 1$.

****

First, calculate $f''(x)$:

$$f(x) = x^2 + \frac{a}{x} = x^2 + ax^{-1}$$

$$f'(x) = 2x - ax^{-2}$$

$$f''(x) = 2 + 2ax^{-3} = 2\left(1 + \frac{a}{x^3}\right)$$

If $f(x)$ has an inflection point at $x = 1$, then $f''(1) = 0$

$$f''(1) = 2\left(1 + \frac{a}{1}\right) = 2(1 + a) = 0 \text{ when } \underline{a = -1}$$

Check: For $a = -1$, $f''(x) = 2\left(1 - \frac{1}{x^3}\right)$

$$\begin{cases} f''(x) < 0 \text{ when } \frac{1}{x^3} > 1 \text{ or } 0 < x < 1 \\ f''(x) > 0 \text{ when } x < 0 \text{ or } x > 1 \end{cases}$$

Hence $f(x)$ is concave <u>down</u> for $x < 1$

$f(x)$ is concave <u>up</u> for $x > 1$

Thus: $x = 1$ is an inflection point of $f(x) = x^2 + \frac{1}{x}$

4-58

Determine concavity for $f(x) = \frac{4}{x^2-1}$.

****

$$f'(x) = \frac{-4(2x)}{(x^2-1)^2} = \frac{-8x}{(x^2-1)^2}$$

$$f''(x) = \frac{(x^2-1)^2(-8) - (-8x)\cdot 2(x^2-1)(2x)}{(x^2-1)^4}$$

$$= \frac{(x^2-1)(-8) - (-8x)\cdot 2\cdot 2x}{(x^2-1)^3} = \frac{24x^2+8}{(x^2-1)^3}.$$

Now the graph of $f(x) = \frac{4}{x^2-1}$ is concave up if and only if $f''(x) > 0$. Since the numerator, $24x^2+8$, is always positive, $f''(x) > 0$ exactly when $x^2-1 > 0$. This is true when $x > 1$ or when $x < -1$. The graph of $f(x)$ is concave up for all x less than -1, is concave down for all x between -1 and $+1$, and is concave up for all x greater than 1.

4-59

Find the values a,b,c so the function $f(x) = x^3 + ax^2 + bx + c$ has a critical point at (1,5) and an inflection point at (2,3).

****

$$f'(x) = 3x^2 + 2ax + b \qquad f''(x) = 6x + 2a$$

$$f''(2) = 0 = 6(2) + 2a \rightarrow a = -6$$

$$f'(1) = 0 = 3(1)^2 + 2(-6)(1) + b \rightarrow b = 9$$

$$f(1) = 5 = (1)^3 + (-6)(1)^2 + 9(1) + C \rightarrow C = 1$$

4-60

Given the function $f(x) = 2x^6 + 9x^5 + 10x^4 - 13x - 5$, determine all intervals on which the graph of f is concave up, all intervals where it is concave down, and find all inflection points for f.

****

$$f'(x) = 12x^5 + 45x^4 + 40x^3 - 13$$

$$f''(x) = 60x^4 + 180x^3 + 120x^2$$

$$= 60x^2(x^2 + 3x + 2)$$

$$= 60x^2(x+1)(x+2)$$

DERIVATIVE SIGN CHART

x		−2		−1		0	
f′	- - - irrelevant - - - - - -						
f″	∪ +	0	∩ −	0	∪ +	0	∪ +
f		21		11		−5	

$f''(1) > 0$

$f''(-1/2) > 0$

$f''(-3/2) < 0$

$f''(-3) > 0$

The graph of f(x) is:

concave up on the intervals $(-\infty, -2)$ and $(-1, \infty)$.

concave down on the interval $(-2, -1)$.

The points $(-2, 21)$ and $(-1, 11)$ are inflection points, while $(0, -5)$ is <u>not</u> an inflection point, since the second derivative does not change sign there.

NEWTON'S METHOD FOR ROOTS OF EQUATIONS

4-61

Use Newton's method to approximate a solution to the following equation.

$$x^3 + 2x = 3.1$$

****

Let $f(x) = x^3 + 2x - 3.1$

$f'(x) = 3x^2 + 2$

$f(x_{n+1}) - f(x_n) \approx f'(x_n)(x_{n+1} - x_n)$ and we want $f(x_{n+1})$ to be at least approximately 0.

Then $x_{n+1} \approx \dfrac{f'(x_n)\,x_n - f(x_n)}{f'(x_n)} = \dfrac{2x_n^3 + 3.1}{3x_n^2 + 2}$

$x_0 = 1$ by inspection of the given equation.

$x_1 = \dfrac{5.1}{5} = 1.02$ is an approximate solution.

4-62

Use Newton's algorithm to determine a root (to two decimal places) of the equation $x^3 + 8x - 23 = 0$, given an initial starting value of $x_o = 2$.

****

Newton's algorithm: $X_{n+1} = X_n - \dfrac{f(x_n)}{f'(x_n)}$

In this problem, we seek a zero of the function $f(x) = x^3 + 8x - 23$. Since $f'(x) = 3x^2 + 8$, then we have:

$$x_{n+1} = x_n - \frac{x_n^3 + 8x_n - 23}{3x_n^2 + 8}$$

$$x_0 = 2$$

$$x_1 = 2 - \frac{8 + 16 - 23}{12 + 8} = 2 - \frac{1}{20} = 1.95$$

$$x_2 = 1.95 - \frac{1.95^3 + 8(1.95) - 23}{3(1.95)^2 + 8} \approx 1.9492$$

$\therefore$ $x = 1.95$ is a root of the given equation, to two decimal places.

4-63

Use Newton's Method to find the root of $6x^3 + x^2 - 19x + 6 = 0$ that lies between 0 and 1.

****

Set $f(x) = 6x^3 + x^2 - 19x + 6$

$f'(x) = 18x^2 + 2x - 19$

$$x = x_0 - \frac{f(x_0)}{f'(x_0)} = x_0 - \frac{6x_0^3 + x_0^2 - 19x_0 + 6}{18x_0^2 + 2x_0 - 19}$$

Begin with $x_0 = 0$:

$$x = 0 - \frac{6}{-19} = \frac{6}{19} = 0.316$$

Now, let $x_0 = 0.316$:

$$x = 0.316 - \frac{6(.316)^3 + (.316)^2 - 19(.316) + 6}{18(.316)^2 + 2(.316) - 19}$$

$$= 0.316 - \frac{0.285}{-16.571} = 0.316 + 0.017 = 0.333$$

Let $x_0 = 0.333$:

$$x = 0.333 - \frac{6(.333)^3 + (.333)^2 - 19(.333) + 6}{18(.333)^2 + 2(.333) - 19}$$

$$= 0.333 - \frac{.0054}{-16.338} = 0.333 + 0.0003 = 0.333$$

The root is 0.333.

4-64

A rule for approximating cube roots states that $\sqrt[3]{a} \approx x_{n+1}$ where

$$x_{n+1} = \frac{1}{3}(2x_n + \frac{a}{x_n^2}) \qquad \text{for } n = 1, 2, 3, \ldots$$

and x_1 is any approximation to $\sqrt[3]{a}$.

a) Use Newton's Method to derive this rule.

b) Use the rule to approximate $\sqrt[3]{9}$.

****

a)

Consider the function: $f(x) = x^3 - a$

Note that $f(x) = 0 \Rightarrow x = \sqrt[3]{a}$

Now apply Newton's Method to approximate the zero of $f(x)$.

Since $f'(x) = 3x^2$ we have:

$$x_{n+1} = x_n - \frac{x_n^3 - a}{3x_n^2} = \frac{3x_n^3 - x_n^3 + a}{3x_n^2} = \frac{1}{3}\left(2x_n + \frac{a}{x_n^2}\right)$$

b)

To approximate $\sqrt[3]{9}$, choose $x_1 = 2$

then $x_2 = \frac{1}{3}(4 + \frac{9}{4}) \approx 2.0833333$

$x_3 \approx \underline{\underline{2.0800886}}$

Which is accurate to 5 decimal places.

GRAPH SKETCHING OF A FUNCTION

4-65

For the curve $f(x) = x^2/(x-2)^2$, find the domain and range, all asymptotes, intervals increasing, intervals concave up, local maximums, local minimums, inflection points, and sketch the curve.

****

$f(x) = \dfrac{x^2}{(x-2)^2}$ $(x-2)^2 = 0$ at $x = 2$, vertical asymptote

$\lim_{x \to +\infty} f(x) = 1 \to y = 1$, horizontal asymptote

$f'(x) = \dfrac{-4x}{(x-2)^3}$ $\quad - \; 0 \; + \; 2 \; -$ local minimum $(0,0)$

local maximum none

increasing $(0,2)$

$f''(x) = \dfrac{8(x+1)}{(x-2)^4}$ $\quad - \; -1 \; + \; 2 \; +$ inflection point $(-1, \frac{1}{9})$

concave up $(-1,2)$ $(2,+\infty)$

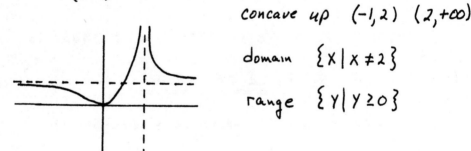

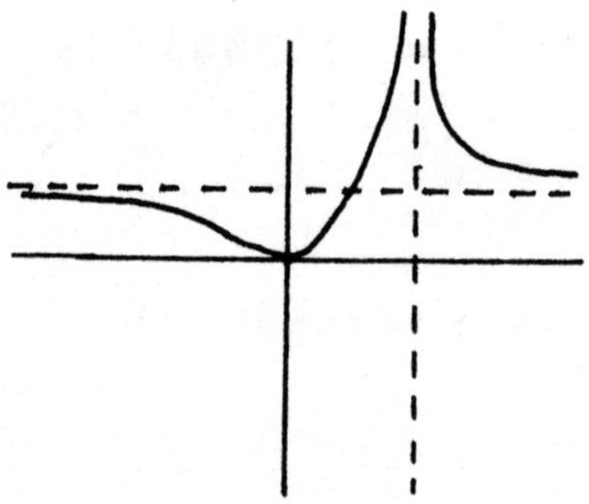

domain $\{x \mid x \neq 2\}$

range $\{y \mid y \geq 0\}$

4-66

Sketch a graph of $y = (2x)/(x^2 + 1)$. Identify and label all asymptotes, extreme points and points of inflection.

****

asymptotes: vertical: none because denominator $\neq 0$ for any real values of x

horizontal: deg num < deg den. so x-axis is horiz. asymp.

oblique: none because deg num < deg den.

extreme values:

$$y' = \frac{(x^2+1)(2) - (2x)(2x)}{(x^2+1)^2} = \frac{2(1-x^2)}{(x^2+1)^2}$$

$$y'' = \frac{(x^2+1)^2(-4x) - (2-2x^2)(2)(x^2+1)(2x)}{(x^2+1)^4} = \frac{-4x(3-x^2)}{(x^2+1)^3}$$

$y' = 0 \Rightarrow 1 - x^2 = 0 \Rightarrow x = \pm 1$

by 2nd deriv test: $y''(1) = -1 < 0$ so $x = 1$ is rel. max.

$y''(-1) = 1 > 0$ so $x = -1$ is rel min.

y' is defined everywhere and we have no domain endpoints so we have no other possible relative extremes

points of inflection: $y'' = 0 \Rightarrow x = 0,\ 3 - x^2 = 0 \Rightarrow x = \pm\sqrt{3}$

since y'' changes sign as it passes through these points we have three points of inflection

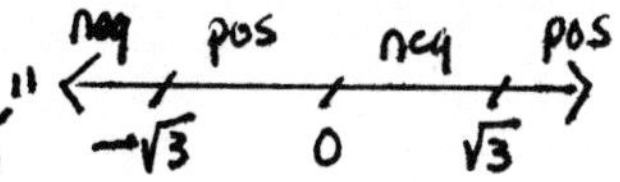

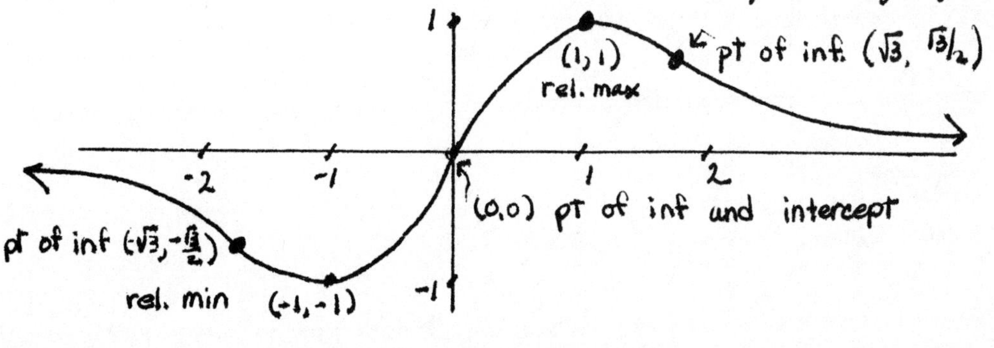

4-67

From the following information about the function f, sketch the graph of f:

Domain: all real numbers except -5.

The first derivative, f′, is such that:

f′(x) > 0 whenever -5 < x < -3 or x > 0;
f′(x) < 0 whenever x < -5 or -3 < x < 0;
f′(x) = 0 at the points (-3,1) and (0,-5).

The second derivative, f", is such that:

f"(x) > 0 whenever -1 < x < 2;
f"(x) < 0 whenever x < -5 or -5 < x < -1 or x > 2;
f"(x) = 0 at the points (-1,-3) and (2,-2).

x-intercepts: (-6,0), (-4,0) and (-2,0).

y-intercept: (0,-5).

$\lim_{x\to\infty} f(x) = 0$ $\lim_{x\to-\infty} f(x) = 2$ $\lim_{x\to-5} f(x) = -\infty$.

****

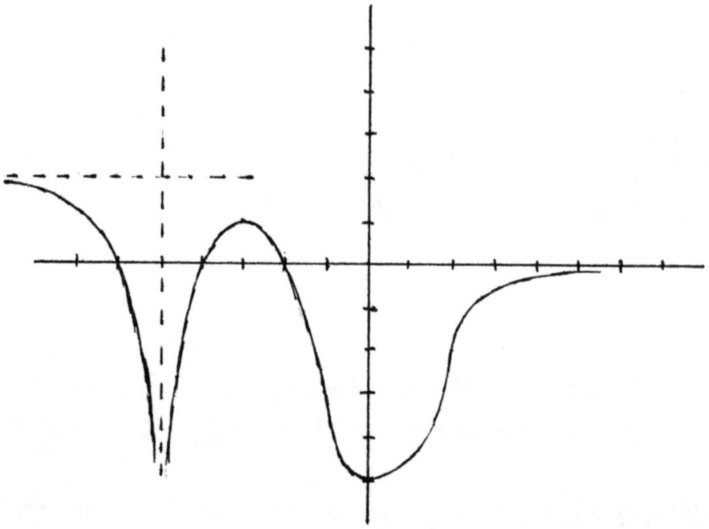

4-68

For the function $f(x) = x^3 - x^2 - 1$, $x \in [-2,3]$, (1) find all relative, endpoint, and absolute maxima and minima, (2) find all points of inflection, (3) find all horizontal and vertical asymptotes, (4) find where f is increasing and where it is decreasing, (5) find where f is concave upward and where it is concave downward, (6) sketch the graph of f.

****

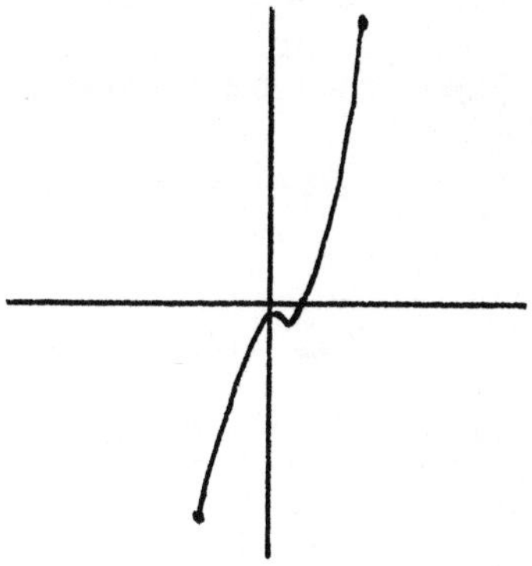

$f'(x) = 3x^2 - 2x$

$f''(x) = 6x - 2$

$f'''(x) = 6$

$3x^2 - 2x = 0$

$x = 0$ OR $x = \frac{2}{3}$

$f''(0) = -$

THUS, RELATIVE MAXIMUM OF -1 AT $x = 0$

$f''(\frac{2}{3}) = +$

THUS, RELATIVE MINIMUM OF $-\frac{31}{27}$ AT $x = \frac{2}{3}$

$6x - 2 = 0$

$x = \frac{1}{3}$

$f'''(\frac{1}{3}) = +$

THUS, POINT OF INFLECTION AT $(\frac{1}{3}, -\frac{29}{27})$

FROM THE GRAPH:

ENDPOINT MAXIMUM OF 17 AT $x = 3$
ENDPOINT MINIMUM OF -13 AT $x = -2$
ABSOLUTE MAXIMUM OF 17 AT $x = 3$
ABSOLUTE MINIMUM OF -13 AT $x = -2$
NO HORIZONTAL OR VERTICAL ASYMPTOTES
INCREASING ON $[-2,0] \cup [\frac{2}{3},3]$
DECREASING ON $[0,\frac{2}{3})$
CONCAVE UP ON $(\frac{1}{3},3]$
CONCAVE DOWN ON $[-2,\frac{1}{3})$

4-69

Consider the function $f(x) = \frac{1}{3}x^3 - \frac{1}{2}x^2 - 2x + 2$.

(a) Find the critical numbers.

(b) Find the interval(s) in which f is increasing.

(c) Find the interval(s) in which f is decreasing.

(d) Find the interval(s) in which the graph of f is concave upward.

(e) Find the interval(s) in which the graph of f is concave downward.

(f) Find the local extrema.

(g) Find any points of inflection.

(h) Sketch the graph of f.

****

$f'(x) = x^2 - x - 2$

$f''(x) = 2x - 1$

(a) SET $f'(x) = 0$. $\therefore (x-2)(x+1) = 0$
HENCE $x = -1, 2$ CRITICAL NUMBERS

(b) $f'(x) \geq 0$ FOR $(-\infty, -1] \cup [2, \infty)$
INCREASING INTERVALS

(c) $f'(x) \leq 0$ FOR $[-1, 2]$ DECREASING INTERVAL

(d) $f''(x) > 0$ FOR $(\frac{1}{2}, \infty)$ CONCAVE UPWARD

(e) $f''(x) < 0$ FOR $(-\infty, \frac{1}{2})$ CONCAVE DOWNWARD

MISCELLANEOUS PROBLEMS

4-28

The vertical face of a dam is the lower half of an ellipse whose major axis is 60 feet long and forms the top of the dam. The minor axis is 40 feet long. Find the force exerted on the face of the dam if the water level is at the top of the dam.

****

$$\frac{x^2}{400} + \frac{y^2}{900} = 1$$

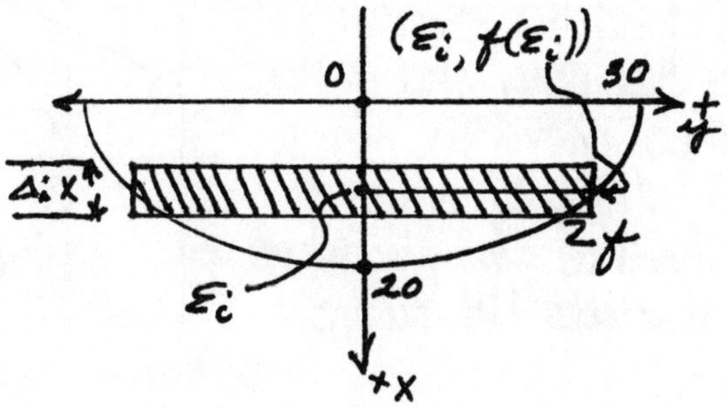

$$9x^2 + 4y^2 = 3600$$

$$y = \frac{3}{2}\sqrt{400 - x^2}$$

The force, F, caused by the liquid pressure is

$$F = \lim_{\|\Delta\| \to 0} \sum_{i=1}^{n} 2\rho\, \varepsilon_i f(\varepsilon_i) \Delta_i x = 2\int_0^{20} \rho\, x f(x)\, dx$$

where ρ lb/ft³ is the weight density of the liquid.

$$F = 2\rho \int_0^{20} x \cdot \frac{3}{2}\sqrt{400 - x^2}\, dx$$

$$= 3\rho \int_0^{20} x\,(400 - x^2)^{1/2}\, dx$$

$$= \frac{3\rho}{-2}\,(400 - x^2)^{3/2} \cdot \frac{2}{3}\Big|_0^{20}$$

$$= -\rho\,(400 - x^2)^{3/2}\Big|_0^{20} = -\rho\,[0 - (8000)]$$

$$= 8000\rho \text{ lbs.}$$

4-29

Show that the tangents drawn from the ends of the focal chord through (4,4) to the parabola $y^2 = 4x$ meet at right angles on the directrix

****

The parabola $y^2=4x$ has vertex $(0,0)$ and focus $(1,0)$. The directrix is the line $x=-1$

To find the co-ordinates of the other end of the focal chord we need to find the equation of the line and find where it meets the curve.

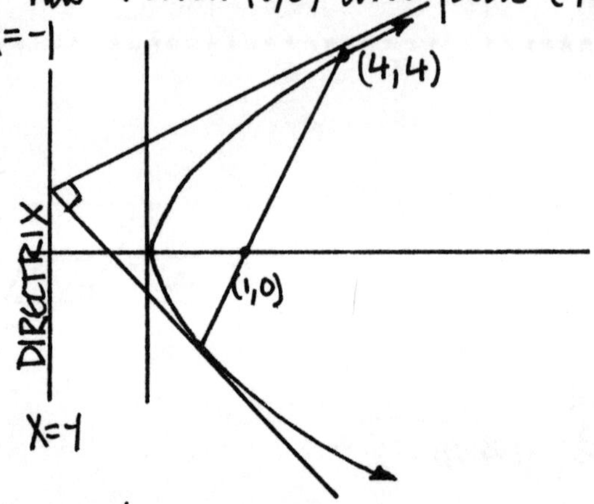

using the point-slope formula. $(y-y_1) = m(x-x_1)$

$$(y-0) = \frac{4-0}{4-1}(x-1) \rightarrow y = \frac{4}{3}x - \frac{4}{3}$$

This line meets the curve again at the solution of the system. $\begin{array}{l} y^2=4x \\ y=\frac{4}{3}x-\frac{4}{3} \end{array}$

let $x = y^2/4$ from $y^2=4x$

so $y = \frac{4}{3}\left(\frac{y^2}{4}\right) - \frac{4}{3} \rightarrow y = \frac{y^2}{3} - \frac{4}{3}$ or $y^2-3y-4=0 \rightarrow (y-4)(y+1)=0$

So the co-ordinate of the second intercept is $y=-1, x=1/4$.

We now need the tangents to $y^2=4x$ at $(4,4)$ and $(1/4,-1)$

To find the slope of the parabola $\frac{d}{dx}(y^2=4x) \rightarrow 2y\frac{dy}{dx} = 4$

That is $\frac{dy}{dx} = \frac{4}{2y}$. So the slopes of the tangents are respectively

$m = 1/2$ and $m = -2$ [notice these lines are perpendicular]

so the equations are

$(y-4) = \frac{1}{2}(x-4)$ $\quad\quad$ $(y - {}^{-}1) = -2(x - 1/4)$

$\Rightarrow 2y - 8 = x - 4$ $\quad\quad$ $\Rightarrow y + 1 = -2x + 1/2$

$\Rightarrow 2y = x + 4$ $\quad\quad$ $\Rightarrow y + 2x + 1/2 = 0.$

Where do the lines meet? from the first substitute $x = 2y - 4$

$y + 2(2y-4) + 1/2 = 0 \rightarrow 5y - 8 + 1/2 = 0$ or $y = 3/2$

now when $y = 3/2$, $x = -1$

Hence the two focal chord tangents are perpendicular and meet on the directrix

4-30

Suppose a conic has a focus at $F = (2,4)$ and the corresponding directrix is $x = 4$. If $P = (6,1)$ is on the conic then the eccentricity of the conic is (a) 1 (b) 3/4 (c) 5/2 (d) 4/5 (e) 5/4 .

****

$$e = \frac{d(P,F)}{d(P,L)} = \frac{5}{2}.$$

4-31

If the following equation is the equation of an ellipse, find its center, foci, major and minor axes, and sketch the graph of the solution.
If the equation is that of a hperbola, find its center (point of symmetry), foci, vertices, the equations of its asymptotes and sketch the graph.

$$2x^2 + 8y^2 - 8x - 16y + 7 = 0.$$

****

Since the coefficients of x^2, y^2 are both positive, the graph of the solution of the equation is an **ellipse**

To find the center and lengths of axes:

$$2x^2 - 8x + 8y^2 - 16y = -7$$

$$\Rightarrow 2(x^2 - 4x + 4) + 8(y^2 - 2y + 1) = -7 + 8 + 8$$

$$\Rightarrow 2(x-2)^2 + 8(y-1)^2 = 9$$

$$\Rightarrow \frac{(x-2)^2}{9/2} + \frac{(y-1)}{9/8} = 9$$

$\therefore$ The center is $(2, 1)$

the major axis is parallel to the x-axis and has length $2\left(\sqrt{9/2}\right) = \frac{6}{\sqrt{2}}$

the minor axis has length $2\left(\sqrt{9/8}\right) = 3/\sqrt{2}$

The foci are located on the major axis at length

$$\sqrt{9/2 - 9/8} = \sqrt{27/8}$$

from the center.

Thus the foci are

$$\left(2 + \sqrt{\tfrac{27}{8}}, 1\right) \text{ and } \left(2 - \sqrt{\tfrac{27}{8}}, 1\right)$$

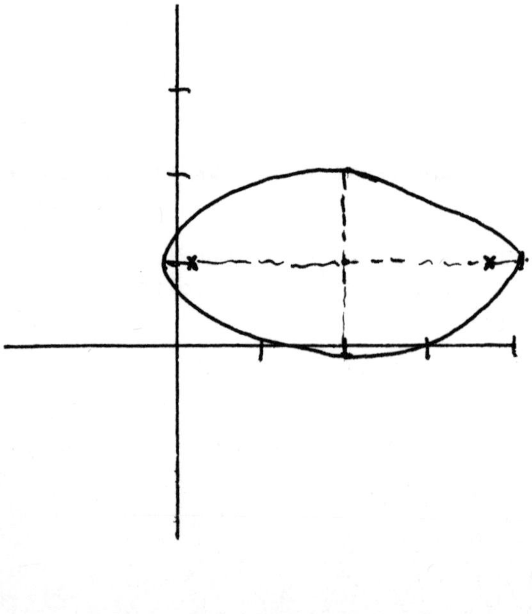

5
POLAR COORDINATES

THE POLAR COORDINATE SYSTEM

5-1

Express the polar equation r = -4 sin θ in rectangular form.

****

$$r = -4 \sin \theta$$
$$r^2 = -4 r \sin \theta$$
$$x^2 + y^2 = -4y$$
$$x^2 + y^2 + 4y = 0$$
$$x^2 + y^2 + 4y + 4 = 4$$
$$x^2 + (y+2)^2 = 4,$$

A CIRCLE OF RADIUS 2 CENTERED AT (0, -2).

NOTE: WHEN WE MULTIPLIED BOTH SIDES OF THE EQUATION BY r, THE RESULTING EQUATION MIGHT HAVE HAD A SOLUTION ($r = 0$) WHICH THE ORIGINAL EQUATION DID NOT HAVE. SINCE $r = 0$ WAS A SOLUTION (WHEN $\theta = 0$) OF THE ORIGINAL EQUATION, WE DID NOT CHANGE THE SOLUTION SET BY MULTIPLYING BY r.

5-2

Let P have rectangular coordinates (1,1). Find four sets of polar coordinates (r,θ) for P such that:

(a) $r > 0$, $\theta > 0$ (b) $r > 0$, $\theta < 0$ (c) $r < 0$, $\theta > 0$, (d) $r < 0$, $\theta < 0$

****

a) $(\sqrt{2}, \pi/4)$

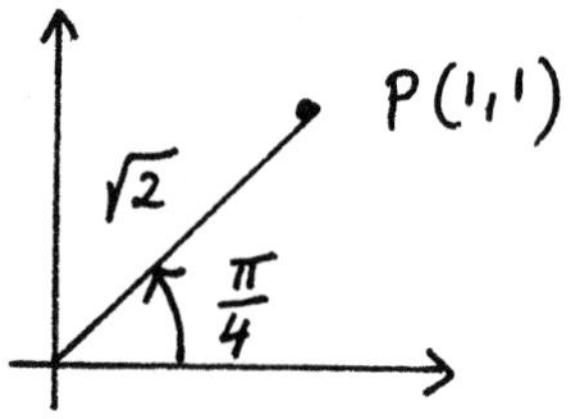

b) $\left(\sqrt{2}, -\frac{7\pi}{4}\right)$

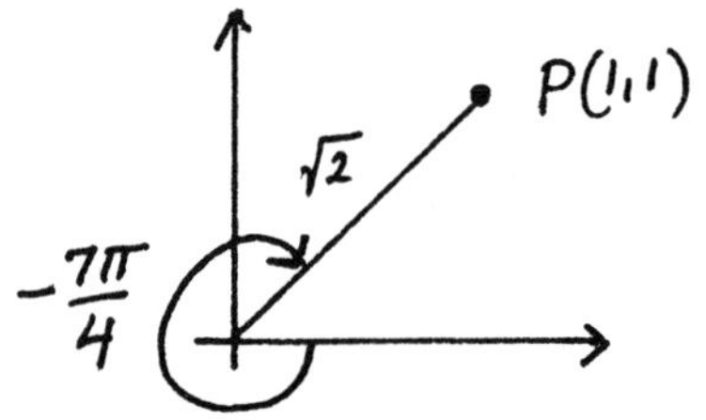

c) $\left(-\sqrt{2}, \frac{5\pi}{4}\right)$

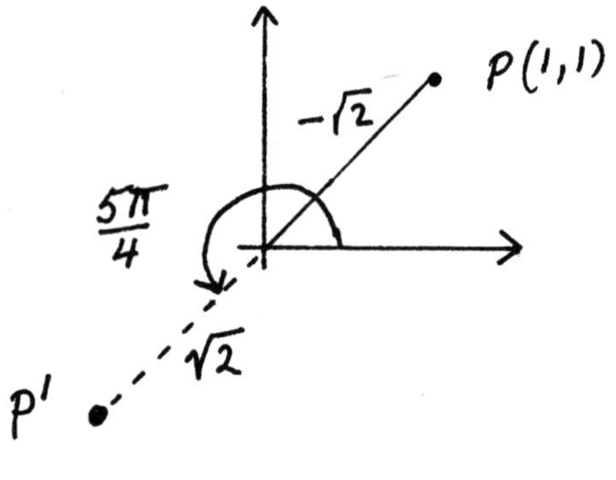

d) $\left(-\sqrt{2}, -\frac{3\pi}{4}\right)$

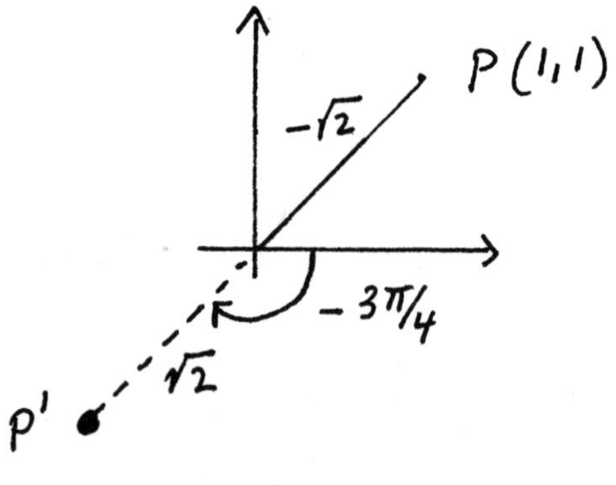

5-3

In each part, do the coordinate conversion requested:

(a) Find polar coordinates for the cartesian point $(-2, 2\sqrt{3})$.

(b) Find cartesian coordinates for the point whose polar coordinates are $(-3, (3\pi/4))$.

(c) Find polar coordinates for the cartesian point $(0,-5)$.

****

(a) $\tan\theta = \frac{y}{x} = -\sqrt{3}$ and the point is in the 2nd quadrant $\therefore\ \theta = \frac{2\pi}{3}$

$r = \sqrt{x^2 + y^2} = \sqrt{4 + 12} = 4$

So, $(r, \theta) = (4, \frac{2\pi}{3})$

(b) $x = r\cos\theta = -3\cos\left(\frac{3\pi}{4}\right) = 3/\sqrt{2}$

$y = r\sin\theta = -3\sin\left(\frac{3\pi}{4}\right) = -3/\sqrt{2}$

$\therefore\ (x, y) = (3/\sqrt{2}, -3/\sqrt{2})$

(c) The point is on the negative y-axis, so $\theta = \frac{3\pi}{2}$ and $(r, \theta) = (5, \frac{3\pi}{2})$

5-4

Convert $x^3 + xy^2 - y^2 = 0$ to polar form and write r explicitly in terms of θ.

****

$$(r\cos\theta)^3 + (r\cos\theta)(r\sin\theta)^2 - (r\sin\theta)^2 = 0$$

$$r^3\cos^3\theta + r^3\cos\theta\sin^2\theta - r^2\sin^2\theta = 0$$

$$r^3\cos\theta(\cos^2\theta + \sin^2\theta) - r^2\sin^2\theta = 0$$

$$r^3\cos\theta - r^2\sin^2\theta = 0$$

$$r\cos\theta - \sin^2\theta = 0 \longrightarrow r = \frac{\sin^2\theta}{\cos\theta} \text{ or } \sin\theta\tan\theta$$

GRAPHING IN POLAR COORDINATES

5-5

Consider the sketch of $r = \sin(4\theta)$ where the leaves are labeled counter clockwise, and number one is the leaf which is symmetric about $\theta = \frac{\pi}{8}$. Find the order that the leaves are sketched as θ moves from 0 to 2π.

****

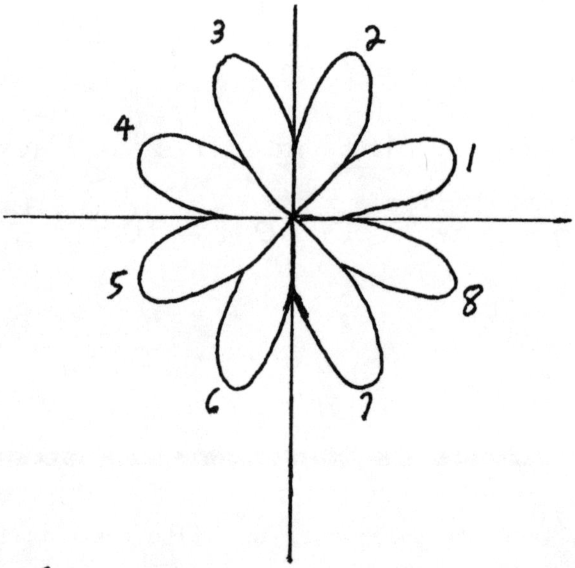

The order is 1-6-3-8-5-2-7-4.

5-6

Compute the eccentricity of the hyperbola whose equation in polar coordinates is

$$r^2 \sin 2\theta = 4$$

by (a) writing the equation in rectangular coordinates, (b) rotating the axes, and (c) computing the eccentricity.

****

(a)

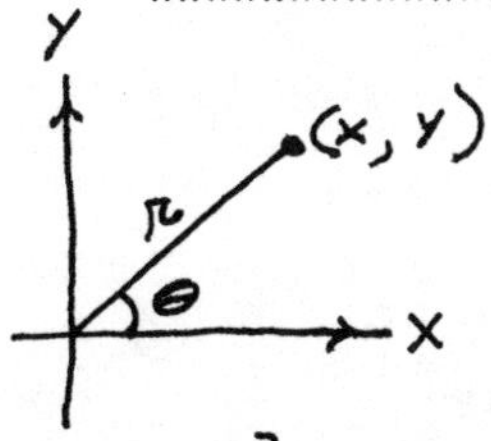

$$\sin 2\theta = 2 \sin\theta \cos\theta = 2\,\frac{y}{r}\,\frac{x}{r}$$

$$r^2 \sin 2\theta = r^2 \cdot \frac{2yx}{r^2} = 4$$

$$xy = 2$$

(b)

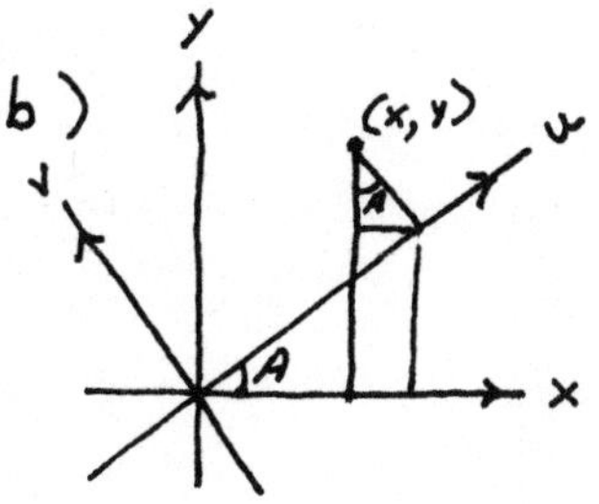

$$x = u\cos A - v\sin A$$

$$y = u\sin A + v\cos A$$

$$xy = (u^2 - v^2)\cos A \sin A + uv(\cos^2 A - \sin^2 A) = 2$$

Let $\cos^2 A - \sin^2 A = 0$. Then $A = \pi/4$.

$$(u^2 - v^2)\frac{1}{\sqrt{2}}\cdot\frac{1}{\sqrt{2}} = 2$$

$\dfrac{u^2}{4} - \dfrac{v^2}{4} = 1$ is in the form $\dfrac{u^2}{a^2} - \dfrac{v^2}{b^2} = 1$

(c) $e = \dfrac{c}{a}$. $a^2 + b^2 = c^2$. $a^2 = b^2 = 4$

$$c^2 = 8$$

$$c = 2\sqrt{2}$$

$$e = \frac{2\sqrt{2}}{2} = \sqrt{2}$$

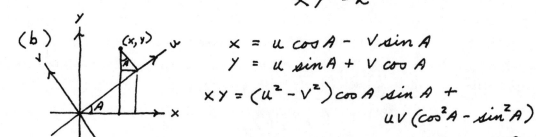

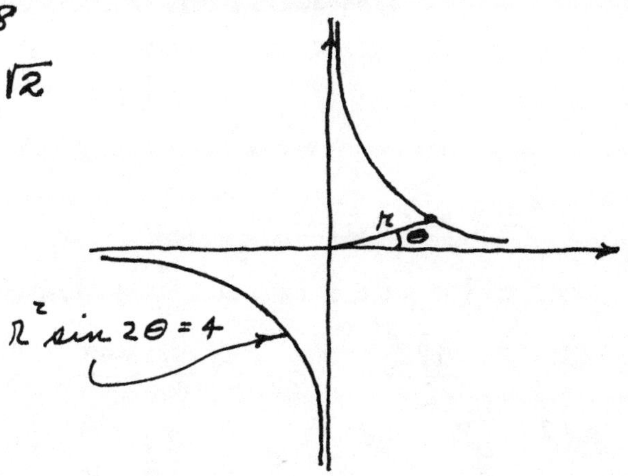

5-7

Find the slope of the line which is tangent to the curve given by the equation $r = 1 + \cos\theta$ at $\theta = \pi/6$.

****

$r = 1 + \cos\theta$

$r' = -\sin\theta$

at $\theta = \pi/6$

$\sin\theta = 1/2$ $\quad r(\pi/6) = 1 + \sqrt{3}/2$

$\cos\theta = \sqrt{3}/2$ $\quad r'(\pi/6) = -1/2$

$$\frac{dy}{dx} = \frac{r'\sin\theta + r\cos\theta}{r'\cos\theta + (-r)\sin\theta}$$

at $\theta = \pi/6$, the slope of the tangent line

$$m = \frac{dy}{dx} = \frac{(-1/2)(1/2) + (1 + \sqrt{3}/2)(\sqrt{3}/2)}{(-1/2)(\sqrt{3}/2) - (1 + \sqrt{3}/2)(1/2)}$$

$$= \frac{-1/4 + \sqrt{3}/2 + 3/4}{-\sqrt{3}/4 - 1/2 - \sqrt{3}/4} = \frac{\sqrt{3}/2 + 1/2}{-\sqrt{3}/2 - 1/2}$$

$$= \boxed{-1}$$

5-8

Graph $r = 3\cos(\theta - \frac{\pi}{4})$.

****

$$r^2 = 3r\cos\left(\theta - \frac{\pi}{4}\right)$$

$$r^2 = 3r\left(\cos\theta\cos\frac{\pi}{4} + \sin\theta\sin\frac{\pi}{4}\right)$$

$$r^2 = \frac{3\sqrt{2}}{2} r\cos\theta + \frac{3\sqrt{2}}{2} r\sin\theta$$

$$x^2 + y^2 = \frac{3\sqrt{2}}{2} x + \frac{3\sqrt{2}}{2} y$$

$$x^2 - \frac{3\sqrt{2}}{2}x + \frac{9}{8} + y^2 - \frac{3\sqrt{2}}{2}y + \frac{9}{8} = \frac{9}{4}$$

$$\left(x - \frac{3\sqrt{2}}{4}\right)^2 + \left(y - \frac{3\sqrt{2}}{4}\right)^2 = \frac{9}{4}$$

CIRCLE WITH CENTER $\left(\frac{3\sqrt{2}}{4}, \frac{3\sqrt{2}}{4}\right)$ AND RADIUS $\frac{3}{2}$

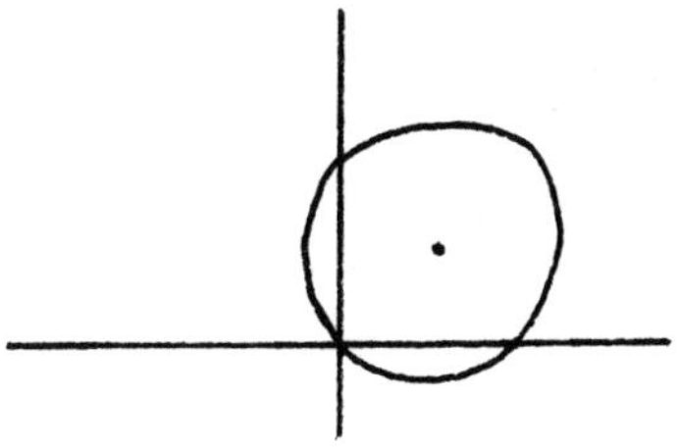

5-9

Discuss the equation $r = 2 - 5\cos\theta$ and draw a sketch of its graph using the polar coordinate system. What will be the effect of making the two numerical values equal to each other ?

****

The given equation is of the general form $r = a \pm b\cos\theta$, where $b > a$. Hence, the graph will be a limaçon in shape with a loop. If $a = b$, the limaçon is known as a cardioid, which looks somewhat similar to a heart.

We also note that we obtain an equivalent equation if the coordinates (r, θ) is replaced by $(r, -\theta)$.

Hence, the graph is symmetric with respect to the polar axis.

The coordinates of some points on the graph is given in the accompanying table, from which half the graph is drawn. Next, using the property of symmetry with respect to the polar axis, the other half is also drawn.

θ	0	$\frac{1}{6}\pi$	$\frac{1}{3}\pi$	$\frac{1}{2}\pi$	$\frac{2}{3}\pi$	$\frac{5}{6}$	π
r	-3	-2.33	-0.5	2	4.5	6.33	7

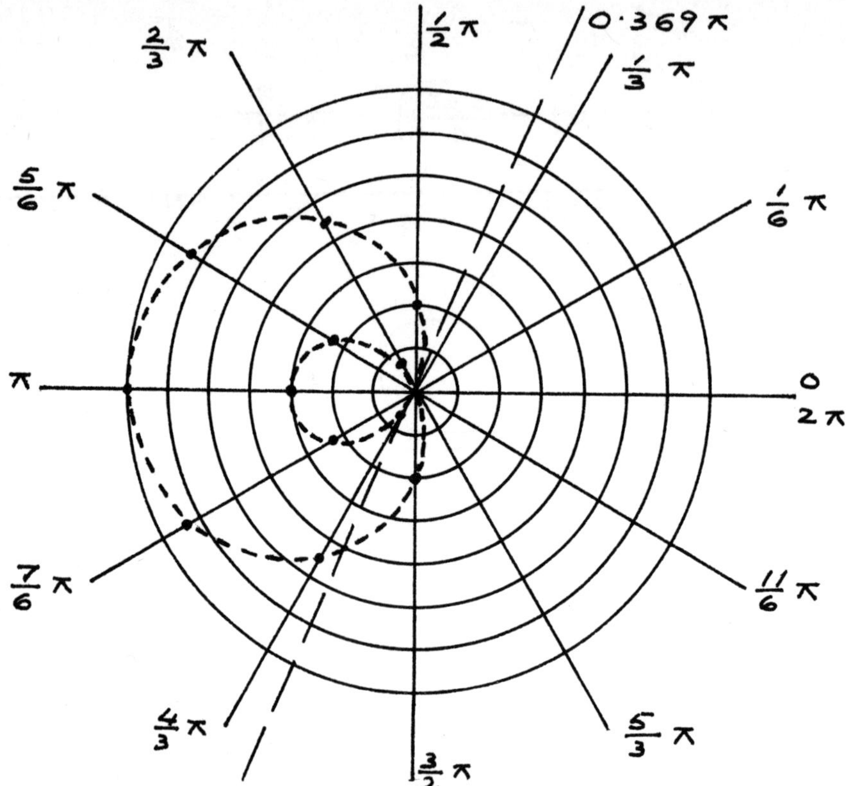

If $r = 0$, we get

$0 = 2 - 5\cos\theta$ OR $\cos\theta = \frac{2}{5}$ and

$\theta = \cos^{-1}\frac{2}{5} = 0.369\pi$ rad

Hence, the point $(0, 0.369\pi)$ is on the graph, and an equation of the tangent line at that point is given by $\theta = 0.369\pi$.

5-10

Sketch the graph of $r = a\cos\theta + a\sin\theta$.

****

Multiply by r, noticing that this will not introduce an extraneous point, to obtain $r^2 = ar\cos\theta + ar\sin\theta$,

$$x^2 + y^2 = ax + ay, \text{ or}$$

$$x^2 - ax + \frac{a^2}{4} + y^2 - ay + \frac{a^2}{4} = \frac{a^2}{2},$$

$$\left(x - \frac{a}{2}\right)^2 + \left(y - \frac{a}{2}\right)^2 = \left(\frac{a}{\sqrt{2}}\right)^2,$$ a circle centered at $\left(\frac{a}{2}, \frac{a}{2}\right)$ of radius $\frac{a}{\sqrt{2}}$.

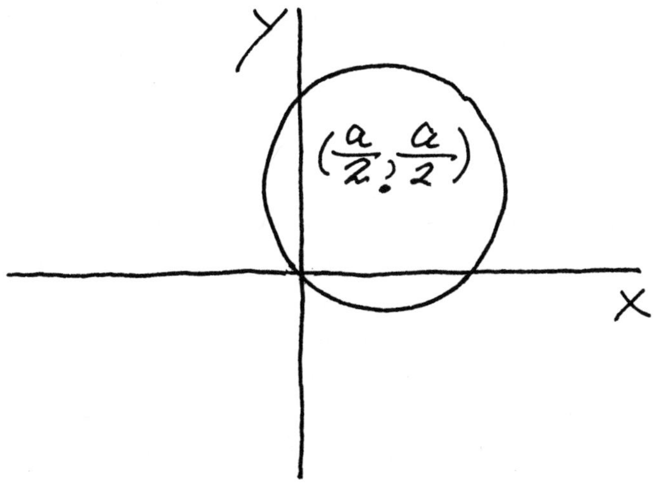

5-11

Graph $r = \frac{1}{2}\theta$ for $0 \leq \theta \leq 4\pi$.

****

θ	r
0	0
$\pi/2$	$\pi/4 \approx .79$
π	$\pi/2 \approx 1.57$
$3\pi/2$	$3\pi/4 \approx 2.36$
2π	$\pi \approx 3.14$
$5\pi/2$	$5\pi/4 \approx 3.93$
3π	$3\pi/2 \approx 4.71$
$7\pi/2$	$7\pi/4 \approx 5.50$
4π	$2\pi \approx 6.28$

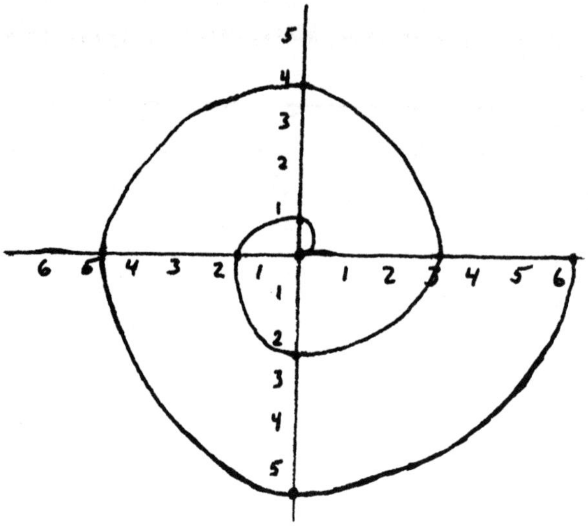

5-12

Sketch $r = \sin^3\left(\frac{\theta}{3}\right)$.

****

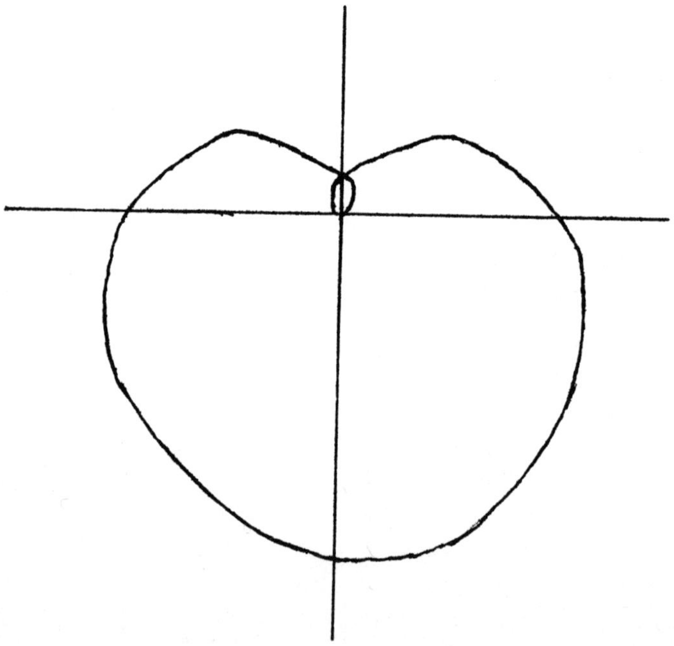

AREA IN POLAR COORDINATES

5-13

Find the area of the region inside the polar curve given by $r = 2 - 2\sin\theta$ and outside the polar curve given by $r = 3$.

****

Find where the cardioid $r = 2 - 2\sin\theta$ and the circle $r = 3$ intersect:

$$3 = 2 - 2\sin\theta, \quad 1 = -2\sin\theta, \quad \sin\theta = \frac{-1}{2}$$

$\therefore$ the region lies between $\theta = \frac{7\pi}{6}$ and $\theta = \frac{11\pi}{6}$.

$$\int \left((2-2\sin\theta)^2 - 9\right) d\theta = \int (4\sin^2\theta - 8\sin\theta - 5)\, d\theta$$

$$= \int \left(2(1-\cos 2\theta) - 8\sin\theta - 5\right) d\theta$$

$$= \int (-2\cos 2\theta - 8\sin\theta - 3)\, d\theta = -\sin 2\theta + 8\cos\theta - 3\theta$$

$$\text{Area} = \frac{1}{2}\int_{\frac{7\pi}{6}}^{\frac{11\pi}{6}} \left((2-2\sin\theta)^2 - 9\right) d\theta$$

$$= \frac{1}{2}\left(-\sin 2\theta + 8\cos\theta - 3\theta\right)\Bigg|_{\frac{7\pi}{6}}^{\frac{11\pi}{6}}$$

$$= \frac{1}{2}\left[\left(\frac{\sqrt{3}}{2} + \frac{8\sqrt{3}}{2} - \frac{11\pi}{2}\right) - \left(\frac{-\sqrt{3}}{2} - \frac{8\sqrt{3}}{2} - \frac{7\pi}{2}\right)\right]$$

$$= \frac{9\sqrt{3}}{2} - \pi$$

5-14

Sketch the graph of $r = 2(\sin\theta + \cos\theta)$.
Find the area of the region enclosed by the curve.

****

a) There is no symmetry with respect to the pole, the polar axis, nor the line $\theta = \pi/2$.

θ	r
0	2
$\pi/6$	$1+\sqrt{3}$
$\pi/4$	$2\sqrt{2}$
$\pi/3$	$1+\sqrt{3}$
$\pi/2$	2
$\frac{3\pi}{4}$	0
π	-2 — completes loop

Note symmetry across $\theta = \pi/4$

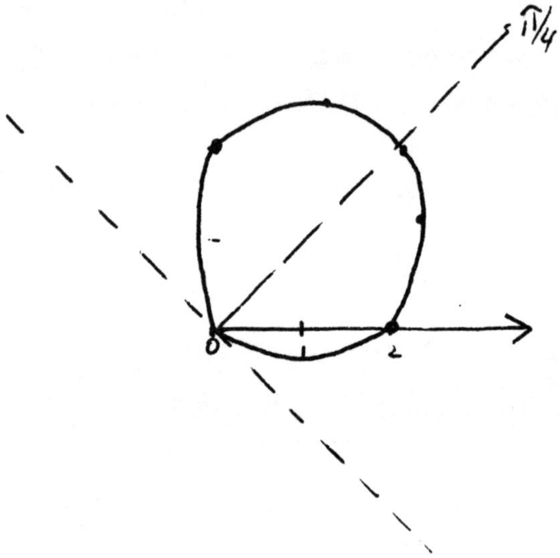

b) $$A = \int_0^{\pi} \frac{1}{2}\left(2(\cos\theta + \sin\theta)\right)^2 \, d\theta$$

$$= \int_0^{\pi} 2\left(\cos^2\theta + \sin^2\theta + 2\sin\theta\cos\theta\right) d\theta$$

$$= \int_0^{\pi} 2 + 2\sin 2\theta \, d\theta$$

← Note: $2\sin\theta\cos\theta = \sin 2\theta$

$$= \left[2\theta + (-\cos 2\theta)\right]\Big|_0^{\pi} = \boxed{2\pi}$$

5-15

a) Sketch the graphs of the circle $r = 6\cos\theta$ and the cardioid $r = 2 + 2\cos\theta$ on the same coordinate system.

b) Find the area of the region in part (a) that is inside the circle and outside the cardioid.

****

a)

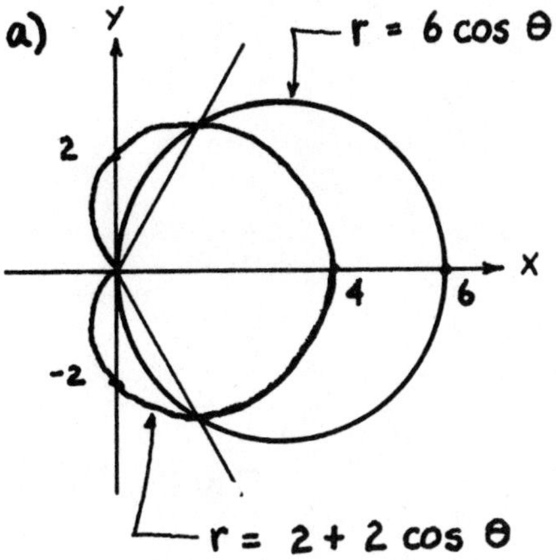

To find the points of intersection, equate the two

$$6\cos\theta = 2 + 2\cos\theta$$

$$\Rightarrow \quad \cos\theta = 1/2$$

$$\Rightarrow \quad \theta = \pi/3 \quad \text{or} \quad 5\pi/3$$

In addition, there is clearly a point of intersection at the origin.

b) Using the area formula $A = \int_\alpha^\beta \frac{1}{2} r^2 \, d\theta$

and taking advantage of symmetry we have

$$A = 2\int_0^{\pi/3} \frac{1}{2}\left[(6\cos\theta)^2 - (2 + 2\cos\theta)^2\right] d\theta$$

$$= \int_0^{\pi/3} (32\cos^2\theta - 8\cos\theta - 4)\, d\theta$$

$$= \left[(16\theta + 8\sin 2\theta) - 8\sin\theta - 4\theta\right]_0^{\pi/3}$$

$$= \underline{\underline{4\pi}}$$

5-16 ■■■

(a)   Sketch the graph of r=2 sin 3θ

(b)   Find the total area outside r=1 and inside r=2 sin 3θ

       ****

(a)

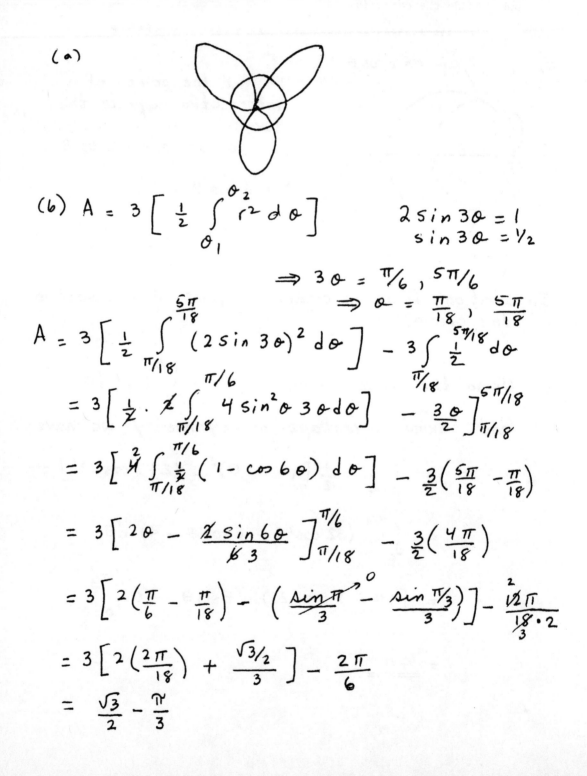

(b) $A = 3\left[\frac{1}{2} \int_{\theta_1}^{\theta_2} r^2 \, d\theta \right]$ $2 \sin 3\theta = 1$
                                                          $\sin 3\theta = \frac{1}{2}$

$$\Rightarrow 3\theta = \frac{\pi}{6}, \frac{5\pi}{6}$$

$$\Rightarrow \theta = \frac{\pi}{18}, \frac{5\pi}{18}$$

$A = 3\left[\frac{1}{2} \int_{\pi/18}^{\frac{5\pi}{18}} (2\sin 3\theta)^2 \, d\theta \right] - 3\int_{\pi/18}^{5\pi/18} \frac{1}{2} \, d\theta$

$= 3\left[\frac{1}{\cancel{2}} \cdot \cancel{4} \int_{\pi/18}^{\pi/6} 4\sin^2 \theta \, 3\theta \, d\theta \right] - \frac{3\theta}{2}\Big]_{\pi/18}^{5\pi/18}$

$= 3\left[\cancel{4}^2 \int_{\pi/18}^{\pi/6} \frac{1}{\cancel{2}} (1-\cos 6\theta) \, d\theta \right] - \frac{3}{2}\left(\frac{5\pi}{18} - \frac{\pi}{18} \right)$

$= 3\left[2\theta - \frac{\cancel{2}\sin 6\theta}{\cancel{6}\,3} \right]_{\pi/18}^{\pi/6} - \frac{3}{2}\left(\frac{4\pi}{18} \right)$

$= 3\left[2\left(\frac{\pi}{6} - \frac{\pi}{18} \right) - \left(\frac{\cancel{\sin\pi}^{\;0}}{3} - \frac{\sin \pi/3}{3} \right) \right] - \frac{\cancel{12}\pi}{\frac{18 \cdot 2}{3}}$

$= 3\left[2\left(\frac{2\pi}{18} \right) + \frac{\sqrt{3}/2}{3} \right] - \frac{2\pi}{6}$

$= \frac{\sqrt{3}}{2} - \frac{\pi}{3}$

5-17

Find the area outside r = 2 but inside r = 4cosθ.

****

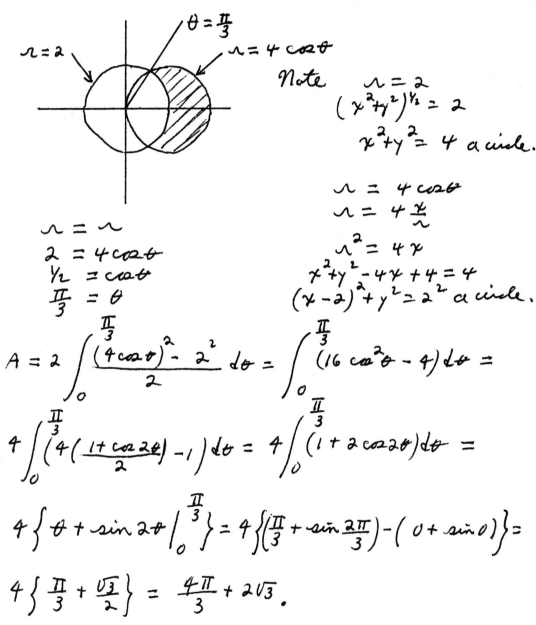

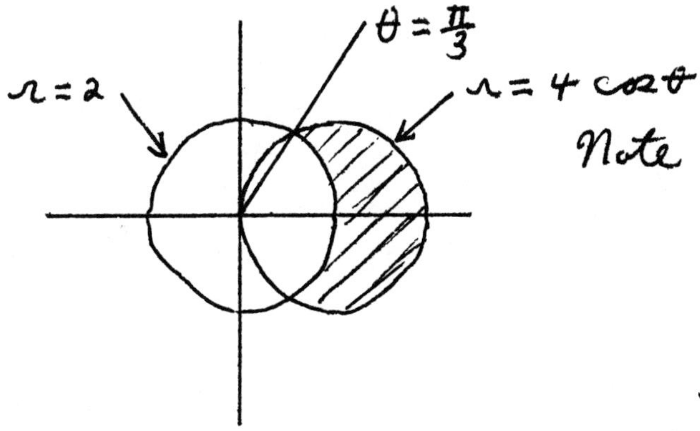

Note

$$r = 2$$
$$(x^2+y^2)^{1/2} = 2$$
$$x^2+y^2 = 4 \text{ a circle.}$$

$$r = 4\cos\theta$$
$$r = 4\frac{x}{r}$$
$$r^2 = 4x$$
$$x^2+y^2-4x+4 = 4$$
$$(x-2)^2+y^2 = 2^2 \text{ a circle.}$$

$$r = r$$
$$2 = 4\cos\theta$$
$$1/2 = \cos\theta$$
$$\frac{\pi}{3} = \theta$$

$$A = 2\int_0^{\frac{\pi}{3}} \frac{(4\cos\theta)^2 - 2^2}{2}\,d\theta = \int_0^{\frac{\pi}{3}} (16\cos^2\theta - 4)\,d\theta =$$

$$4\int_0^{\frac{\pi}{3}} \left(4\left(\frac{1+\cos 2\theta}{2}\right) - 1\right)d\theta = 4\int_0^{\frac{\pi}{3}} (1 + 2\cos 2\theta)\,d\theta =$$

$$4\left\{\theta + \sin 2\theta \Big|_0^{\frac{\pi}{3}}\right\} = 4\left\{\left(\frac{\pi}{3} + \sin\frac{2\pi}{3}\right) - (0 + \sin 0)\right\} =$$

$$4\left\{\frac{\pi}{3} + \frac{\sqrt{3}}{2}\right\} = \frac{4\pi}{3} + 2\sqrt{3}.$$

5-18

Find the area of all possible even leaf roses whose leaf length is one.

****

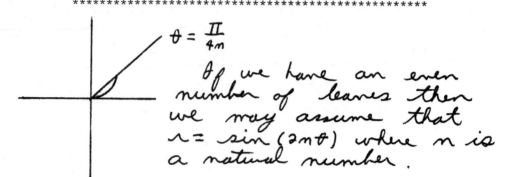

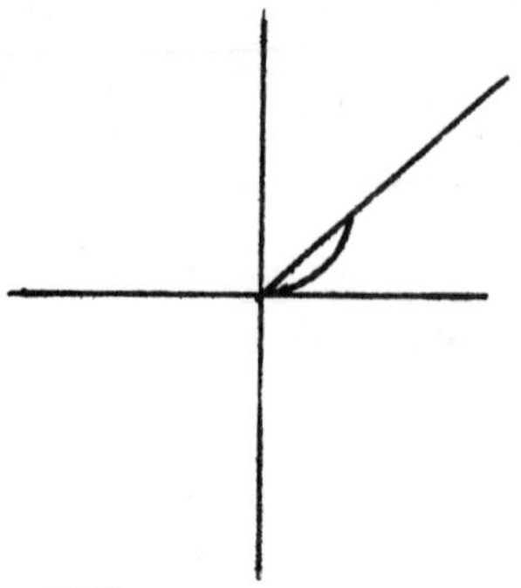

If we have an even number of leaves then we may assume that $r = \sin(2n\theta)$ where n is a natural number.

The rose $r = \sin(2n\theta)$ where n is a natural number has $4n$ leaves. If we find the area of one-half of a leaf and then multiply that area by $8n$ we will have the entire area. From the sketch it is clear that as θ moves from 0 to $\frac{\pi}{4n}$ one-half of a leaf is completed.

The area $= A = 8n \int_0^{\frac{\pi}{4n}} \frac{\sin^2(2n\theta)}{2} d\theta =$

$$4n \int_0^{\frac{\pi}{4n}} \frac{1-\cos(4n\theta)}{2} d\theta = 2n \int_0^{\frac{\pi}{4n}} (1-\cos(4n\theta)) d\theta =$$

$$2n\left\{ \theta - \frac{\sin 4n\theta}{4n} \Big|_0^{\frac{\pi}{4n}} \right\} = 2n\left\{\left(\frac{\pi}{4n} - 0\right) - (0-0)\right\} =$$

$$2n\left(\frac{\pi}{4n}\right) = \frac{\pi}{2}.$$

5-19

Sketched below is the propeller (with equation r = 2 sin 3θ) to be mounted atop the beanies which will soon be required head gear for all mathematics, physics, and chemistry majors. As the proud wearer walks, the propeller will spin causing the words "mathematics", "physics", and "chemistry" printed on the blades to blur and blend signifying the unity of pure science. Find the area of ONE BLADE of the propeller.

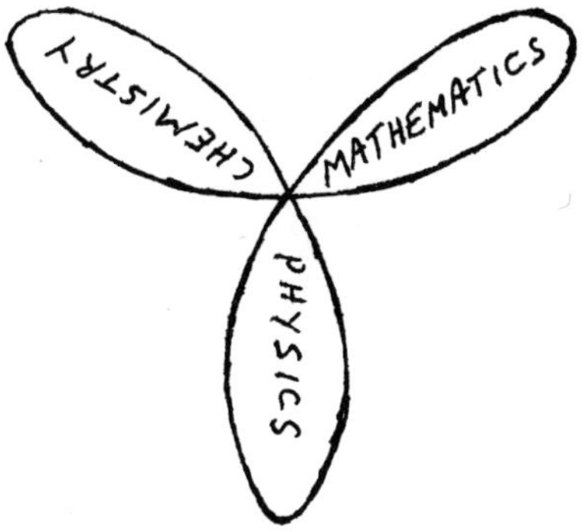

****

SET $r = 0$

$2 \sin 3\theta = 0$

$3\theta = m\pi$, m AN INTEGER

$\theta = \frac{\pi}{3} m$.

SO ONE BLADE IS SWEPT OUT AS θ GOES FROM 0 TO $\frac{\pi}{3}$.

$$\int_0^{\pi/3} \frac{1}{2} r^2 \, d\theta = \int_0^{\pi/3} \frac{1}{2}(2 \sin 3\theta)^2 \, d\theta$$

$$= \int_0^{\frac{\pi}{3}} 2 \sin^2 3\theta \, d\theta$$

$$= \int_0^{\frac{\pi}{3}} 2 \, \frac{1 - \cos 6\theta}{2} \, d\theta$$

$$= \int_0^{\pi/3} (1 - \cos 6\theta) \, d\theta$$

$$= \theta - \frac{1}{6} \sin 6\theta \Big|_0^{\pi/3}$$

$$= \frac{\pi}{3} - \frac{1}{6} \sin 2\pi - \left(0 - \frac{1}{6} \sin 0\right)$$

$$= \frac{\pi}{3}$$

5-20

Find the area enclosed by the three- leaved rose curve p = 2 sin 3θ .

****

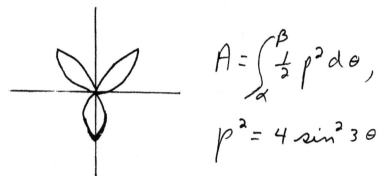

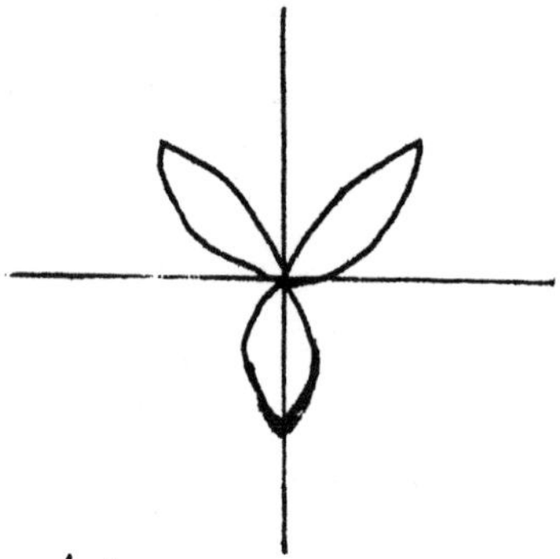

$$A = \int_{\alpha}^{\beta} \frac{1}{2} p^2 \, d\theta,$$

$$p^2 = 4 \sin^2 3\theta$$

We find the area of one of the leaves and multiply by 3 to find the total area.

$$A = 3 \int_0^{\frac{\pi}{3}} \frac{1}{2}(4 \sin^2 3\theta) \, d\theta,$$

$$= 3(2) \int_0^{\frac{\pi}{3}} \sin^2 3\theta \, d\theta$$

$$= \frac{6}{3} \int_0^{\frac{\pi}{3}} 3 \sin^2 3\theta \, d\theta$$

$$= 2 \left[\frac{1}{2}(3\theta) - \frac{1}{4} \sin 6\theta \right]_0^{\frac{\pi}{3}}$$

$$= 3\theta - \frac{1}{2} \sin 6\theta \Big|_0^{\frac{\pi}{3}}$$

$$= \left[\pi - \tfrac{1}{2} 0\right] - [0 - 0]$$

$$= \pi \text{ square units}$$

5-21

Use polar coordinates to write the equation of each circle sketched below. Then find the area of the shaded region.

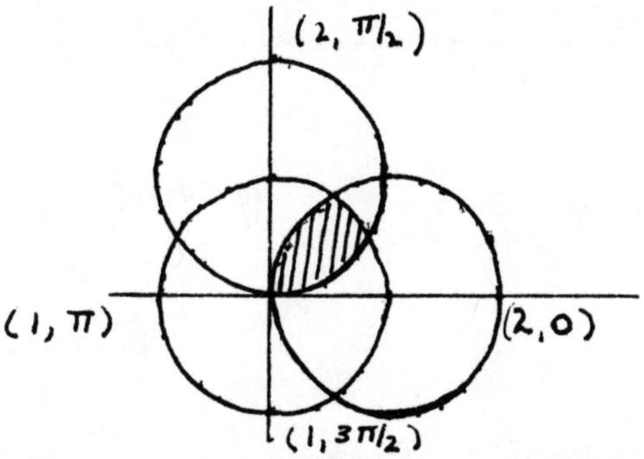

****

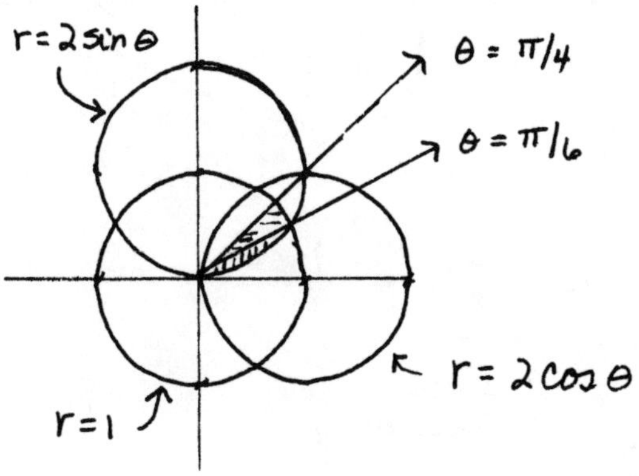

Find points of intersection

$2\cos\theta = 2\sin\theta$

$1 = \frac{\sin\theta}{\cos\theta} = \tan\theta$

$\Rightarrow \theta = \pi/4$

$1 = 2\sin\theta$

$1/2 = \sin\theta \Rightarrow \theta = \pi/6$

note: region is symmetric about $\theta = \pi/4$ so we will find area from $\theta = 0$ to $\theta = \pi/4$, then double it.

The region is bounded by two curves so two integrals are necessary.

$$A = 2\left[\int_0^{\pi/6} \tfrac{1}{2}(2\sin\theta)^2\,d\theta + \int_{\pi/6}^{\pi/4} \tfrac{1}{2}(1)^2\,d\theta\right.$$

$$= \int_0^{\pi/6} 4\sin^2\theta\,d\theta + \int_{\pi/6}^{\pi/4} d\theta$$

$$= 4\int_0^{\pi/6} \frac{1-\cos 2\theta}{2}\,d\theta + \int_{\pi/6}^{\pi/4} d\theta$$

$$= 2\left(\theta - \frac{\sin 2\theta}{2}\right)\Big|_0^{\pi/6} + \theta\Big|_{\pi/6}^{\pi/4}$$

$$= \frac{5\pi - 6\sqrt{3}}{12}$$

5-22

Find the area of the region R inside the circle $r = \sin\theta$ and outside the cardioid $r = 1 + \cos\theta$.

****

R is the hatched region in the figure

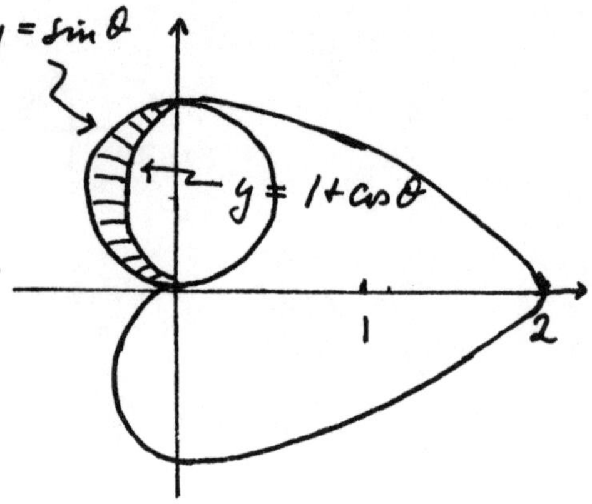

In polar coordinates, area is:

$$A = \frac{1}{2}\int_{\alpha}^{\beta} r^2\, d\theta$$

Hence:

$$A = \frac{1}{2}\int_{\frac{\pi}{2}}^{\pi} (\sin^2\theta - (1+\cos\theta)^2)\, d\theta$$

($\sin^2\theta$: outer boundary; $(1+\cos\theta)^2$: inner boundary)

$$= \frac{1}{2}\int_{\pi/2}^{\pi} (\sin^2\theta - 1 - 2\cos\theta - \cos^2\theta)\, d\theta$$

$$= \frac{1}{2}\int_{\pi}^{\pi/2} (1 + 2\cos\theta + \cos^2\theta - \sin^2\theta)\, d\theta$$

Note: we reversed the limits of integration

$$= \frac{1}{2}\int_{\pi}^{\pi/2} (1 + 2\cos\theta + \cos 2\theta)\, d\theta$$

$$= \frac{1}{2}\left(\theta + 2\sin\theta + \frac{1}{2}\sin 2\theta\right]_{\pi}^{\pi/2}$$

$$= \frac{1}{2}\left[\left(\frac{\pi}{2} + 2\sin\frac{\pi}{2} + \frac{1}{2}\sin\pi\right) - \left(\pi + 2\sin\pi + \frac{1}{2}\sin 2\pi\right)\right]$$

$$= \frac{1}{2}\left[\left(\frac{\pi}{2} + 2 + 0\right) - (\pi + 0 + 0)\right]$$

$$= \frac{1}{2}\left(2 - \frac{\pi}{2}\right) = 1 - \frac{\pi}{4} = \frac{4-\pi}{4}$$

5-23

Calculate the total area enclosed by the leaves of $r = 5\cos 3\theta$

****

Set $5\cos 3\theta = 0$ where $0 \le \theta < 2\pi$.

$3\theta = \frac{m\pi}{2}$ where $m = 1, 3, 5, 7, \cdots$

$\theta = \frac{m\pi}{6} = \frac{\pi}{6}, \frac{\pi}{2}, \frac{5\pi}{6}, \frac{7\pi}{6}, \frac{3\pi}{2}, \frac{11\pi}{6}$

Therefore, there are 6 leaves, but there may be overlapping. Plot the polar graph:

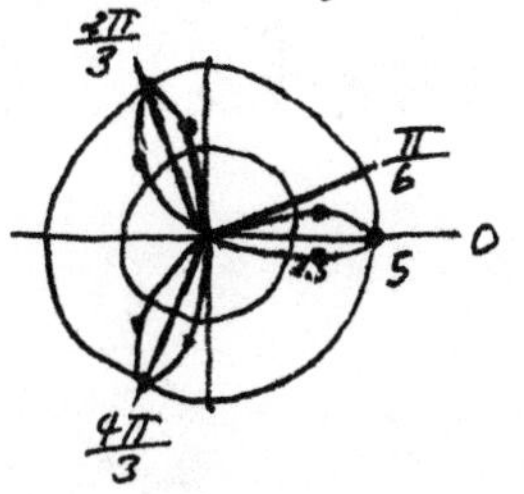

$$A = \frac{1}{2}\int r^2 d\theta = \frac{1}{2}\int (5\cos 3\theta)^2 d\theta$$

$$= 6 \times \frac{1}{2}\int_0^{\frac{\pi}{6}} 25\cos^2 3\theta\, d\theta$$ (There are 3 distinct leaves, or 6 half-leaves.)

$$= 75\int_0^{\frac{\pi}{6}} \frac{1+\cos 6\theta}{2}\, d\theta = \frac{75}{2}\left(\theta + \frac{\sin 6\theta}{6}\right)\Big|_0^{\frac{\pi}{6}}$$

$$= \frac{75}{2}\left(\frac{\pi}{6} - 0 + 0 - 0\right) = \frac{25}{4}\pi \text{ square units.}$$

5-24

Sketch the bifolium $r = a \sin\Theta\cos^2\Theta$. Find the total area enclosed.

****

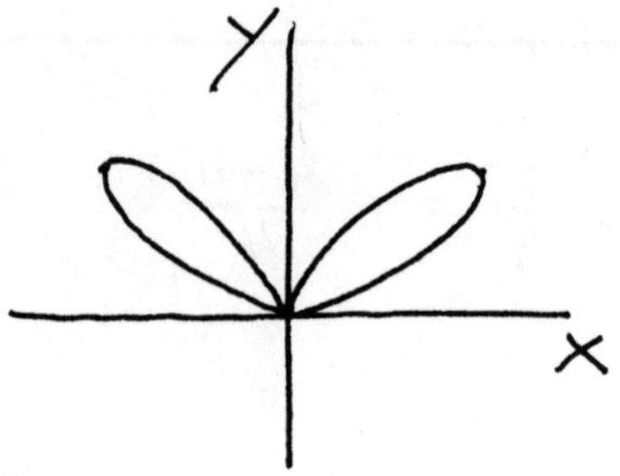

$$A = \int_{\theta_1}^{\theta_2} \frac{1}{2} r^2 d\theta$$

$$= \int_0^{\pi} \frac{1}{2} a^2 \sin^2\theta \cos^4\theta d\theta$$

$$= \frac{1}{2}\int_0^{\pi} a^2\cos^4\theta - a^2\cos^6\theta \, d\theta$$

At this point a reduction formula is useful: $\int \cos^n x\,dx = \frac{\cos^{n-1}x \sin x}{n} + \frac{n-1}{n}\int \cos^{n-2}x\,dx$

$$\int_0^{\pi} \cos^6\theta\,d\theta = \left.\frac{\cos^5\theta\sin\theta}{6}\right|_0^{\pi} + \frac{5}{6}\int_0^{\pi}\cos^4\theta\,d\theta$$

$$= \frac{5}{6}\int_0^{\pi} \cos^4\theta \, d\theta,$$

$$A = \frac{1}{2}a^2 \cdot \frac{1}{6}\int_0^{\pi} \cos^4\theta \, d\theta$$

$$= \frac{a^2}{12}\left[\frac{\cos^3\theta \sin\theta}{4}\Big|_0^{\pi} + \frac{3}{4}\int_0^{\pi} \cos^2\theta \, d\theta\right]$$

$$= \frac{a^2}{16}\int_0^{\pi} \cos^2\theta \, d\theta$$

$$= \frac{a^2}{16}\left[\frac{\cos\theta \sin\theta}{2}\Big|_0^{\pi} + \frac{1}{2}\int_0^{\pi} 1 \, d\theta\right] = \frac{\pi a^2}{32}.$$

5-25

Find the area of one petal of the rose $r = 3 \sin(5\theta)$.

****

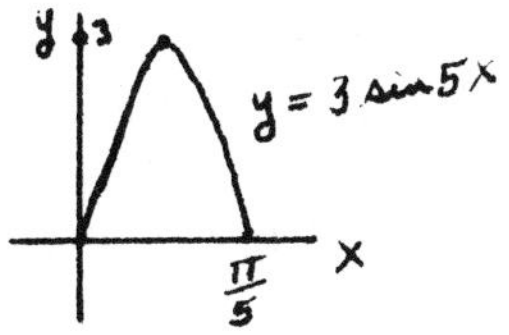

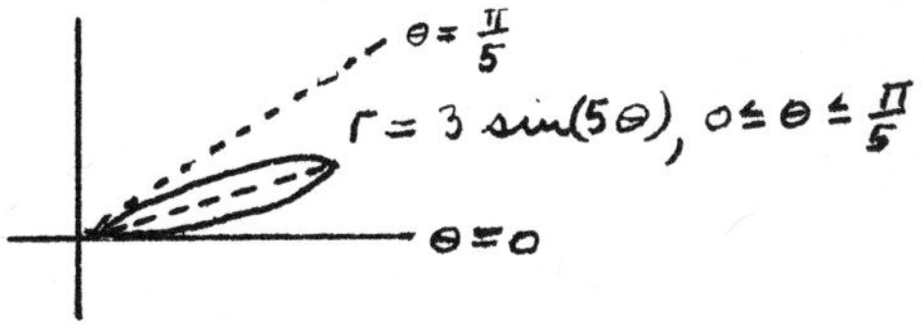

A petal is obtained as 5θ varies from 0 to π, ie. as θ varies from 0 to $\frac{\pi}{5}$. $A = \int_0^{\frac{\pi}{5}} \frac{1}{2} r^2 \, d\theta =$ $\frac{9}{2}\int_0^{\frac{\pi}{5}} \sin^2 5\theta \, d\theta = \frac{9}{2}\int_0^{\frac{\pi}{5}} \frac{1 - \cos 10\theta}{2} \, d\theta =$ $\frac{9}{4}\left[\theta - \frac{1}{10}\sin 10\theta\right]_0^{\frac{\pi}{5}} = \frac{9}{4} \cdot \frac{\pi}{5} = \frac{9\pi}{20}.$

5-26

Find the area of all possible odd leaf roses whose leaf length is one.

****

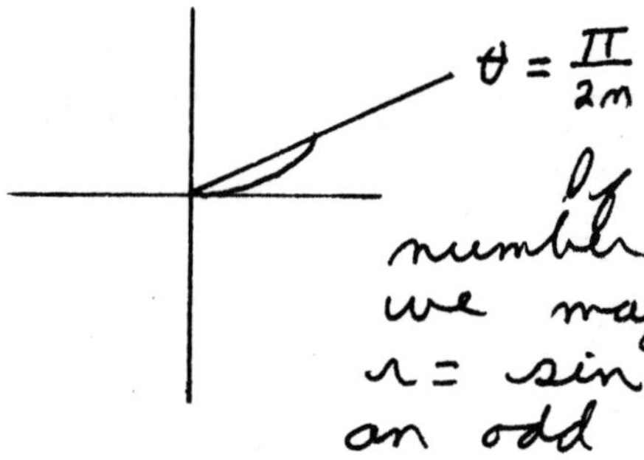

If we have an odd number of leaves then we may assume that $r = \sin(n\theta)$ where n is an odd integer.

The rose $r = \sin(n\theta)$ where n is odd has n leaves. If we find the area of one-half of a leaf and then multiply that area by $2n$ we will have the entire area. From the sketch it is clear that as θ moves from 0 to $\frac{\pi}{2n}$ one-half of a leaf will be completed.

The area $= A = 2n \int_0^{\frac{\pi}{2n}} \frac{1}{2}(\sin nx)^2 dx =$

$$n \int_0^{\frac{\pi}{2n}} \frac{1 - \cos(2nx)}{2} dx = \frac{1}{2} n \int_0^{\frac{\pi}{2n}} (1 - \cos(2nx))\, dx =$$

$$\frac{n}{2} \left\{ x - \frac{\sin(2nx)}{2n} \Big|_0^{\frac{\pi}{2n}} \right\} = \frac{n}{2} \left\{ \left(\frac{\pi}{2n} - 0\right) - (0 - 0) \right\} = \frac{\pi}{4}.$$

5-27

Sketch the graph and compute the area enclosed by the graph of the polar equation

$$r = 2 + \cos 2\theta$$

****

θ	2θ	$\cos 2\theta$	r
0	0	1	3
$\frac{\pi}{4}$	$\frac{\pi}{2}$	0	2
$\frac{\pi}{2}$	π	-1	1
$\frac{3\pi}{4}$	$\frac{3\pi}{2}$	0	2
π	2π	1	3

Since $\cos(-A) = \cos A$, the graph is symmetric with respect to the x-axis

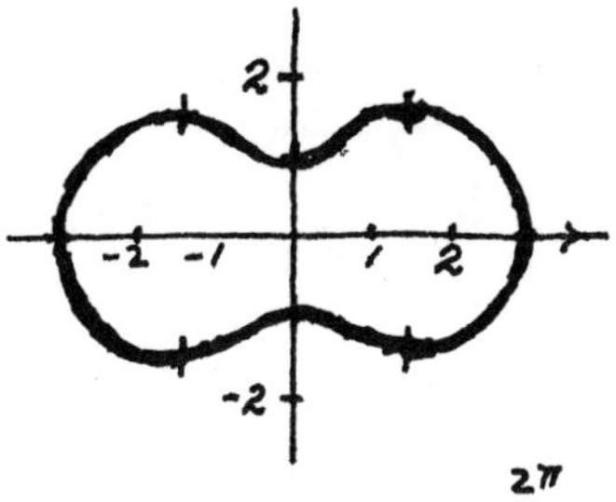

$$A = \frac{1}{2}\int_0^{2\pi} r^2\,d\theta = \frac{1}{2}\int_0^{2\pi}(2+\cos 2\theta)^2\,d\theta = \frac{1}{2}\int_0^{2\pi}(4 + 4\cos 2\theta + \cos^2 2\theta)\,d\theta$$

$$= \frac{1}{2}\left\{\left[4\theta + 2\sin 2\theta\right]_0^{2\pi} + \frac{1}{2}\int_0^{2\pi}(1 + \cos 4\theta)\,d\theta\right.$$

$$= 4\pi + \frac{1}{4}\left[\theta + \frac{1}{4}\sin 4\theta\right]_0^{2\pi} = 4\pi + \frac{\pi}{2} = \frac{9\pi}{2}$$

5-28

Find the area of the region inside the circle r=3 and outside the spiral r=θ, between the values of θ where θ=0 and where the two graphs intersect.

****

$r = \theta$ and $r = 3$ intersect when $\theta = 3$. Thus

$$A = \int_0^3 \frac{1}{2}\left((3)^2 - (\theta)^2\right)d\theta = \frac{1}{2}\int_0^3 (9 - \theta^2)\,d\theta$$

$$= \frac{1}{2}\left[9\theta - \frac{1}{3}\theta^3\right]_0^3 = \frac{1}{2}\left[(27 - 9) - 0\right] = \frac{18}{2} = 9$$

5-29

Find the area inside the curve $r = 4 \cos 2\theta$ and outside the curve $r = 2$.

****

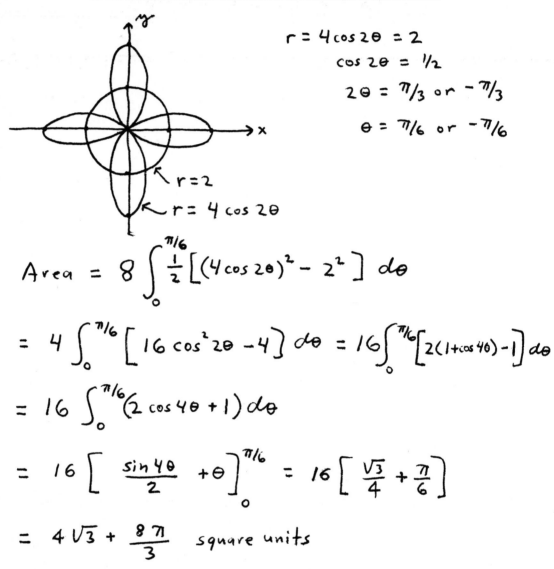

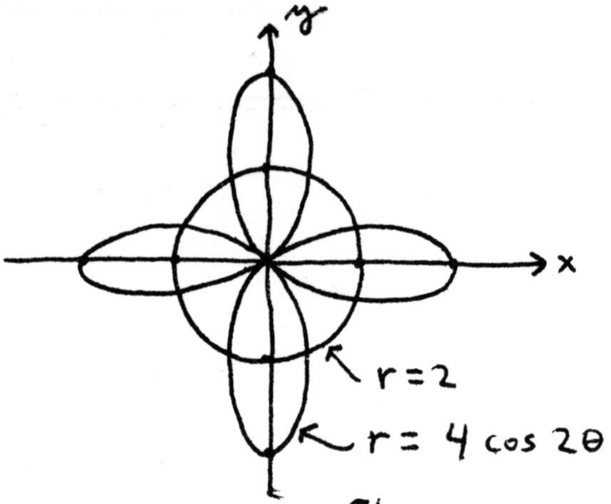

$$r = 4\cos 2\theta = 2$$
$$\cos 2\theta = 1/2$$
$$2\theta = \pi/3 \text{ or } -\pi/3$$
$$\theta = \pi/6 \text{ or } -\pi/6$$

$$\text{Area} = 8\int_0^{\pi/6} \frac{1}{2}\left[(4\cos 2\theta)^2 - 2^2\right] d\theta$$

$$= 4\int_0^{\pi/6}\left[16\cos^2 2\theta - 4\right] d\theta = 16\int_0^{\pi/6}\left[2(1+\cos 4\theta) - 1\right] d\theta$$

$$= 16\int_0^{\pi/6}(2\cos 4\theta + 1)\, d\theta$$

$$= 16\left[\frac{\sin 4\theta}{2} + \theta\right]_0^{\pi/6} = 16\left[\frac{\sqrt{3}}{4} + \frac{\pi}{6}\right]$$

$$= 4\sqrt{3} + \frac{8\pi}{3} \text{ square units}$$

POINTS OF INTERSECTION IN POLAR EQUATIONS

5-30

Find the points of intersection of $r = 3\cos\theta$ and $r = 3 - 3\cos\theta$.

****

$$3\cos\theta = 3 - 3\cos\theta$$

$$\cos\theta = \frac{1}{2}$$

$$\theta = \frac{\pi}{3} \quad \text{OR} \quad \theta = \frac{5\pi}{3}$$

$$\therefore\ r = \frac{3}{2}$$

POINTS OF INTERSECTION ARE $\left[\frac{3}{2}, \frac{\pi}{3}\right], \left[\frac{3}{2}, \frac{5\pi}{3}\right], [0, 0]$

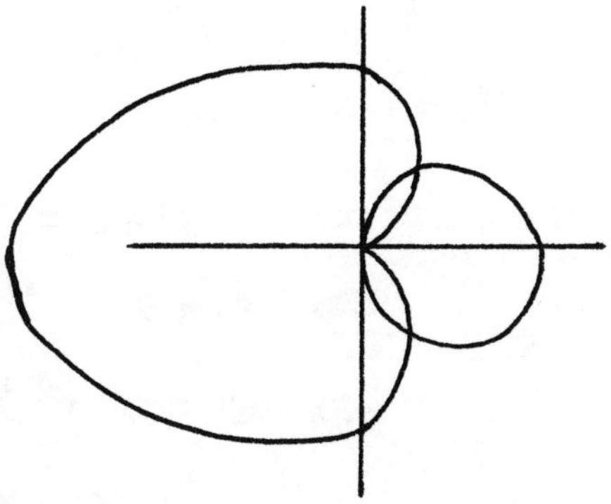

5-31

Find all points of intersection of the curve r = 2 - 4 sinθ with the curve r = 2 sinθ.

****

A rough sketch will save time in estimating the number of points of intersection:

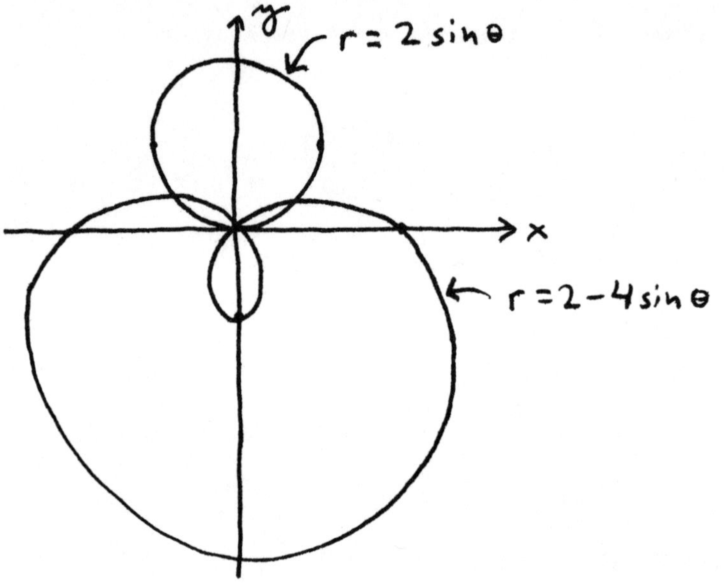

Clearly, the pole is a point of intersection: $(0, \pi/6)$ and $(0,0)$

Also: $2\sin\theta = 2-4\sin\theta \Rightarrow 6\sin\theta = 2$

$\Rightarrow \sin\theta = 1/3$

$\Rightarrow \theta = \sin^{-1} 1/3$

or $\pi - \sin^{-1} 1/3$

Thus, $(2/3, \sin^{-1} 1/3)$ is a second intersection point, and $(2/3, \pi - \sin^{-1} 1/3)$ is the third point of intersection.

6

INDETERMINATE FORMS AND IMPROPER INTEGRALS

L'HÔPITAL'S RULE AND INDETERMINATE FORMS

6-1

Evaluate: $\lim_{x \to +\infty} \dfrac{e^x - e^{2x}}{e^x + e^{-x}}$

****

divide all terms by e^{2x} to get:

$$\lim_{x \to +\infty} \frac{1/e^x - 1}{1/e^x + 1/e^{3x}}$$

as $x \to +\infty$, $e^x \to +\infty$, $e^{3x} \to +\infty$,
so $1/e^x \to 0$, $1/e^{3x} \to 0$

so $\lim_{x \to +\infty} \dfrac{1/e^x - 1}{1/e^x + 1/e^{3x}} \to \dfrac{0 - 1}{0 + 0} \to -\infty$

6-2

Evaluate the following limit, if it exists

$$\lim_{x \to 0} \frac{e^{-2x} - 1}{x^2 - x}$$

****

Since $\lim_{x \to 0} (e^{-2x} - 1) = e^0 - 1 = 1 - 1 = 0$ and

$\lim_{x \to 0} (x^2 - x) = 0 - 0 = 0$, the limit has the indeterminant form $\frac{0}{0}$. Applying L'Hôpital's rule

$$\lim_{x \to 0} \frac{e^{-2x} - 1}{x^2 - x} = \lim_{x \to 0} \frac{-2e^{-2x}}{2x - 1} = \frac{-2e^0}{2(0) - 1}$$

$$= \frac{-2}{-1} = 2$$

6-3

Evaluate the limit: $\lim_{x \to 0} \left(\frac{\sin 6x}{\sin x}\right)^2$

****

$$\lim_{x \to 0} \left(\frac{\sin 6x}{\sin x}\right)^2 = \left(\lim_{x \to 0} \frac{\sin 6x}{\sin x}\right)^2 = \text{form } \frac{0}{0}$$

Then, by L'Hospital's Rule:

$$= \left(\lim_{x \to 0} \frac{6 \cos 6x}{\cos x}\right)^2 = \left(\frac{6 \cdot 1}{1}\right)^2$$

$$= 36$$

Thus, $\lim_{x \to 0} \left(\frac{\sin 6x}{\sin x}\right)^2 = 36.$

6-4

Evaluate the limit $\lim_{x \to 0} \frac{e^{5x}-5x-1}{x^2}$.

****

AS $x \to 0$, $e^{5x}-5x-1 \to 0$ AND $x^2 \to 0$, SO WE CAN APPLY L'HÔPITAL'S RULE

$$\lim_{x \to 0} \frac{e^{5x}-5x-1}{x^2} = \lim_{x \to 0} \frac{5e^{5x}-5}{2x}.$$

BUT, AS $x \to 0$, $5e^{5x}-5 \to 0$ AND $2x \to 0$, SO WE CAN APPLY L'HÔPITAL'S RULE AGAIN

$$\lim_{x \to 0} \frac{5e^{5x}-5}{2x} = \lim_{x \to 0} \frac{25e^{5x}}{2} = \frac{25}{2}.$$

6-5

Evaluate the following limit:

$$\lim_{x \to 0} \frac{2e^x - x^2 - 2x - 2}{x^4 + x^3}$$

****

Using L'Hopital's rule,

$$\lim_{x \to 0} \frac{2e^x-x^2-2x-2}{x^4+x^3} = \lim_{x \to 0} \frac{2e^x-2x-2}{4x^3+3x^2}$$

$$= \lim_{x \to 0} \frac{2e^x-2}{12x^2+6x} = \lim_{x \to 0} \frac{2e^x}{24x+6} = \frac{1}{3}$$

6-6

Find $\lim_{x \to 0} \frac{\ln(1+x)}{x^3}$, if it exists.

****

Both numerator and denominator have the value zero for $x=0$. Also, both have continous derivatives in the neighborhood of $x=0$.

Applying L'Hospital's rule

$$\lim_{x \to 0} \frac{\ln(1+x)}{x^3} = \lim_{x \to 0} \frac{\frac{1}{1+x}}{3x^2}$$

The numerator of the result approaches 1 and the denominator approaches 0 as $x \to 0$.

Therefore, the fraction increases without limit and we conclude that

$$\lim_{x \to 0} \frac{\ln(1+x)}{x^3} \to \infty \quad (\text{DOES NOT EXIST})$$

6-7

Find $\lim_{x \to 0} (1 + \sin \pi x)^{1/x}$.

****

Indeterminate form. Let $\omega = (1 + \sin \pi x)^{1/x}$.

Then $\lim_{x \to 0} (\ln \omega) = \lim_{x \to 0} \dfrac{\ln(1 + \sin \pi x)}{x}$

(now can apply L'Hôpital's rule)

$$= \lim_{x \to 0} \frac{\left(\dfrac{\pi \cos \pi x}{1 + \sin \pi x}\right)}{1} = \pi .$$

$$\lim_{x \to 0} (\ln \omega) = \pi \quad \therefore \quad \lim_{x \to 0} \omega = e^{\pi}$$

6-8

Evaluate: $\lim_{x \to 0^+} x^2(\ln x)$

****

$\lim_{x \to 0^+} x^2(\ln x) \to (0)(-\infty)$ indeterminate form

rewrite: $\lim_{x \to 0^+} \dfrac{\ln x}{x^{-2}} = \lim_{x \to 0^+} \dfrac{\ln x}{(1/x^2)} \to \dfrac{-\infty}{+\infty}$

this is an indeterminate form to which we may apply L'Hôpital's Rule.

$$\lim_{x \to 0^+} \frac{\ln x}{x^{-2}} = \lim_{x \to 0^+} \frac{1/x}{-2x^{-3}} = \lim_{x \to 0^+} \left(-\tfrac{1}{2}\right)x^2 \to 0.$$

6-9

Find $\lim_{x\to\infty}\left(1 - \frac{2}{x}\right)^{3x}$

****

This is of the form 1^{∞}

Let $y = \lim_{x\to\infty}\left(1 - \frac{2}{x}\right)^{3x}$

$$\Rightarrow \ln y = \lim_{x\to\infty} \ln\left(1 - \frac{2}{x}\right)^{3x}$$

$$= \lim_{x\to\infty} 3x \ln\left(1 - \frac{2}{x}\right)$$

$$= \lim_{x\to\infty} \frac{\ln\left(1 - \frac{2}{x}\right)}{\frac{1}{3x}}$$

Using L'Hôpital's Rule:

$$\ln y = \lim_{x\to\infty} \frac{\left(\frac{1}{1-2x}\right)\left(+\frac{2}{x^2}\right)}{-\frac{1}{3x^2}} = -6$$

$$\Rightarrow y = e^{-6}$$

6-10

Find the limit

$$\lim_{x \to 0} \frac{xe^{2x} - x}{1 - \cos x}$$

****

Since $\lim_{x \to 0} (xe^{2x} - x) = \lim_{x \to 0} (1 - \cos x) = 0$

the given limit is an indeterminate form of type $\left(\frac{0}{0}\right)$. Thus by L'Hôpital's Rule

$$\lim_{x \to 0} \frac{xe^{2x} - x}{1 - \cos x} = \lim_{x \to 0} \frac{2xe^{2x} + e^{2x} - 1}{\sin x}$$

But $\lim_{x \to 0} (2xe^{2x} + e^{2x} - 1) = \lim_{x \to 0} \sin x = 0$

and the limit is again an indeterminate form of type $\left(\frac{0}{0}\right)$. By L'Hôpital's Rule

$$\lim_{x \to 0} \frac{2xe^{2x} + e^{2x} - 1}{\sin x} = \lim_{x \to 0} \frac{2(2xe^{2x} + e^{2x}) + 2e^{2x}}{\cos x}$$

$$= \lim_{x \to 0} \frac{4xe^{2x} + 4e^{2x}}{\cos x}$$

$$= \underline{\underline{4.}}$$

6-11

Evaluate $\lim_{x \to 3} \frac{x^n - 3^n}{x^m - 3^m}$ (m ≠ 0)

****

$$\lim_{x \to 3} \frac{x^n - 3^n}{x^m - 3^m} \quad (m \neq 0)$$

$$= \lim_{x \to 3} \frac{n x^{n-1}}{m x^{m-1}} \quad \text{BY L'HOSPITALS RULE}$$

$$= \lim_{x \to 3} \frac{n}{m} x^{n-m} = \frac{n}{m}(3)^{n-m}$$

6-12

Find $\lim_{x \to 0} \left[\frac{e^x - e^{-x}}{\sin x}\right]$.

****

Since $e^x - e^{-x}$ and $\sin x$ both approach 0 as $x \to 0$, the ratio $\frac{e^x - e^{-x}}{\sin x}$ is an indeterminant form of $\frac{0}{0}$; hence, by L'Hospital's Rule we have

$$\lim_{x \to 0} \left[\frac{e^x - e^{-x}}{\sin x}\right] = \lim_{x \to 0} \left[\frac{e^x + e^{-x}}{\cos x}\right]$$

$$= 2$$

6-13

By repeated use of l'hospitals rule evaluate the limit

$$\lim_{x \to 0} \frac{\text{arc sin } x - x}{\sin x - x}$$

****

$$\lim_{x \to 0} \frac{\arcsin x - x}{\sin x - x} = \frac{0-0}{0-0} \cdot \text{Type } \frac{0}{0}$$ The limit satisfies the condition for l'hopitals rule

applying l'hopital's rule. $\lim_{x \to a} \frac{f(x)}{g(x)} = \lim_{x \to a} \frac{f'(x)}{g'(x)}$

$$\lim_{x \to 0} \frac{\arcsin x - x}{\sin x - x} = \lim_{x \to 0} \frac{\frac{1}{\sqrt{1-x^2}} - 1}{\cos x - 1}$$

evaluate $\lim_{x \to 0} \frac{\frac{1}{\sqrt{1-x^2}} - 1}{\cos x - 1} = \frac{1-1}{1-1} = \frac{0}{0}$

So applying l'hopital's rule again

$$\lim_{x \to 0} \frac{(1-x^2)^{-\frac{1}{2}} - 1}{\cos x - 1} = \lim_{x \to 0} \frac{-\frac{1}{2}(1-x^2)^{-\frac{3}{2}}(-2x)}{-\sin x} = \lim_{x \to 0} \frac{\frac{x}{(1-x^2)^{3/2}}}{-\sin x} = \frac{0}{0}$$

so we must apply l'hopitals rule for a third time

$$\lim_{x \to 0} \frac{x(1-x^2)^{-3/2}}{-\sin x} = \lim_{x \to 0} \frac{x\left[\frac{3}{2}(1-x^2)^{-5/2}(-2x)\right] + (1-x^2)^{-3/2}}{-\cos x}$$

$$= \lim_{x \to 0} \frac{\frac{-3x^2}{(1-x^2)^{5/2}} + \frac{1}{(1-x^2)^{3/2}}}{-\cos x} = \frac{0+1}{-1} = -1$$

Hence by three applications of l'hopitals rule

$$\lim_{x \to 0} \frac{\arcsin x - x}{\sin x - x} = -1$$

6-14

Find the limit of f(x) = (x + cosx)/x as x approaches infinity.

****

Rewrite $\frac{X+\cos X}{X} = \frac{X}{X} + \frac{\cos X}{X} = 1 + \frac{\cos X}{X}$

Then $\lim_{X\to\infty} f(x) = \lim_{X\to\infty}\left(1 + \frac{\cos X}{X}\right) = 1 + \lim_{X\to\infty}\frac{\cos X}{X}$. By intuition the last limit is 0 and can be verified by the "squeeze theorem." Hence $\lim_{X\to\infty} f(x) = 1 + 0 = \underline{\underline{1}}$

NOTE: If we try to apply L'Hospital's Rule for the ∞/∞ form we get the incorrect result "no limit." Proceeding blindly yields:

$$\lim_{X\to\infty} f(x) = \lim_{X\to\infty}\frac{1-\sin x}{1} = 1 - \lim_{X\to\infty}\sin x .$$ Since

sinx oscillates endlessly between -1 and 1 as X "gets large", the conclusion would be no limit. The trouble is that L'Hospital's Rule is valid only if $\lim_{X\to\infty}\frac{g'}{f'}$ exists. It does not! (For us $\frac{g'(x)}{f'(x)} = \frac{1-\sin x}{1} = 1-\sin x$.)

6-15

Find the limit of f(t) = (1-cost) / $\sin^2 t$ as t approaches zero.

****

By the L'Hospital Rule for the "0/0" form,

$$\lim_{t\to 0} f(t) = \lim_{t\to 0}\frac{\sin t}{2\sin t\cos t} = \lim_{t\to 0}\frac{1}{2\cos t} = \frac{1}{2}.$$ ANS.

6-16

Evaluate:

$$\lim_{x \to 0} \left(\frac{1}{x} + \log x\right) \qquad \text{(the base of the logarithm is e)}$$

****

$\lim_{x \to 0} \left(\frac{1}{x} + \log x\right)$ is of the form $\infty - \infty$

Change to: $\lim_{x \to 0} \left[\frac{1 + x \log x}{x}\right]$ which has an indeterminate form in the numerator.

We thus evaluate $\lim_{x \to 0} [x \log x]$

$$= \lim_{x \to 0} \left[\frac{\log x}{1/x}\right] \quad \text{which has form } \frac{\infty}{\infty}$$

by l'Hôpital's rule $= \lim_{x \to 0} \left[\frac{1/x}{-1/x^2}\right]$

$$= \lim_{x \to 0} [-x] = 0$$

thus $\lim_{x \to 0} \left[\frac{1 + x \log x}{x}\right]$ is of the form $\frac{1}{0}$

and therefore **the limit does not exist.**

6-17

Evaluate the following limit if it exists.

$$\lim_{x\to 0^+} x^{2x}$$

****

$\lim_{X\to 0^+} X^{2X}$ $[= 0^0]$ ∴ indeterminate.

LET $y = x^{2x}$.

HENCE, $\ln y = 2X \ln X$.

$\lim_{X\to 0^+} \ln y = \lim_{X\to 0^+} 2X \ln X$ $[= 0(-\infty)]$ ∴ INDETERMINATE.

$= \lim_{X\to 0^+} \dfrac{2\ln X}{\frac{1}{X}}$, $\left[= \dfrac{-\infty}{\infty}\right]$ ∴ APPLY L'HÔPITAL'S RULE.

$= \lim_{X\to 0^+} \dfrac{\frac{2}{X}}{-\frac{1}{X^2}} = \lim_{X\to 0^+} (-2X) = 0$

∴ SINCE $\ln X^{2X} = 2X \ln X$, THEN $\lim_{X\to 0^+} \ln X^{2X} = 0$.

BUT $\lim_{X\to 0^+} \ln X^{2X} = \ln \lim_{X\to 0^+} X^{2X} = 0$

∴ $\lim_{X\to 0^+} X^{2X} = 1$

6-18

Evaluate the limit : $\lim_{x \uparrow \pi/2} (\tan x)^{\cos x}$

****

Let $y = (\tan x)^{\cos x}$

Hence, by taking logarithms on both sides, we have

$$\ln y = \cos x \cdot \ln \tan x$$

$$= \frac{1}{\sec x} \cdot \ln \tan x$$

Also, $\lim_{x \uparrow \pi/2} \ln y = \lim_{x \uparrow \pi/2} \dfrac{\dfrac{\sec^2 x}{\tan x}}{\sec x \cdot \tan x}$

$$= \lim_{x \uparrow \pi/2} \frac{\sec^2 x}{\sec x} \cdot \frac{1}{\tan^2 x}$$

$$= \lim_{x \uparrow \pi/2} \frac{1}{\cos x} \cdot \frac{\cos^2 x}{\sin^2 x}$$

$$= \lim_{x \uparrow \pi/2} \frac{\cos x}{\sin^2 x}$$

$$= 0$$

Hence, we write

$$\ln \left(\lim_{x \uparrow \pi/2} y \right) = 0$$

and $\lim_{x \uparrow \pi/2} (\tan x)^{\cos x} = 1$ <u>Ans.</u>

6-19

Find the following limit (if it exists): $\lim\limits_{x\to\infty} (1+x^2)^{1/x}$.

****

Let $y = (1+x^2)^{1/x}$ then $\ln y = \frac{1}{x}\ln(1+x^2)$

Using L'Hospital's Rule on $\ln y$ gives:

$$\lim_{x\to\infty} \frac{\ln(1+x^2)}{x} = \lim_{x\to\infty} \frac{2x}{1+x^2} = \lim_{x\to\infty} \frac{2}{2x}$$

$$= \lim_{x\to\infty} \frac{1}{x} = 0$$

Since $\lim\limits_{x\to\infty} \ln y = 0$, $\lim\limits_{x\to\infty} y = \lim\limits_{x\to\infty} (1+x^2)^{1/x} = e^0$

$= 1.$

6-20

Evaluate the following limits: (a) $\lim\limits_{x\to 1} \dfrac{1 - x + \ln x}{x^3 - 3x + 2}$ (b) $\lim\limits_{x\to\infty} x^3/e^x$

(c) $\lim\limits_{x\to 0} (1 + \sinh x)^{3/x}$.

****

(a) $\lim\limits_{x\to 1} \dfrac{1-x+\ln x}{x^3-3x+2} \swarrow \frac{0}{0} = \lim\limits_{x\to 1} \dfrac{-1+\frac{1}{x}}{3x^2-3} \swarrow \frac{0}{0} = \lim\limits_{x\to 1} \dfrac{-\frac{1}{x^2}}{6x} = \dfrac{-1}{6}.$

(b) $\lim\limits_{x\to\infty} \dfrac{x^3}{e^x} \swarrow \frac{\infty}{\infty} = \lim\limits_{x\to\infty} \dfrac{3x^2}{e^x} \swarrow \frac{\infty}{\infty} = \lim\limits_{x\to\infty} \dfrac{6x}{e^x} \swarrow \frac{\infty}{\infty} = \lim\limits_{x\to\infty} \dfrac{6}{e^x} = 0$

(c) $(1+\sinh x)^{3/x} = e^{\frac{3\ln(1+\sinh x)}{x}}$ and

$\lim\limits_{x\to 0} \dfrac{3\ln(1+\sinh x)}{x} \swarrow \frac{0}{0} = \lim\limits_{x\to 0} \dfrac{3\cosh x}{1+\sinh x} = 3$, therefore

$\lim\limits_{x\to 0} (1+\sinh x)^{3/x} = e^3.$

6-21

Find $\lim_{x \to 0} \frac{x - \sin x}{x^3}$.

****

This is an indeterminate form $\frac{0}{0}$.

$$\lim_{x \to 0} \frac{x - \sin x}{x^3} = \lim_{x \to 0} \frac{1 - \cos x}{3x^2}$$

$$= \lim_{x \to 0} \frac{\sin x}{6x}$$

$$= \lim_{x \to 0} \frac{\cos x}{6} = \frac{1}{6}.$$

6-22

Evalute the following limit if it exists.

$$\lim_{x \to 0} \frac{\ln \sec^4 x}{x^2}$$

****

$$\lim_{x \to 0} \frac{\ln \sec^4 x}{x^2} = \lim_{x \to 0} \frac{4 \ln \sec x}{x^2} \quad \left[= \frac{0}{0}\right] \therefore \text{INDETERMINATE.}$$

APPLY L'HÔPITAL'S RULE.

$$= 4 \lim_{x \to 0} \frac{\frac{1}{\sec x} \sec x \tan x}{2x}$$

$$= 4 \lim_{x \to 0} \frac{\tan x}{2x} \quad \left[= \frac{0}{0}\right]$$

$\therefore$ APPLY L'HÔPITAL'S RULE.

$$= 4 \lim_{x \to 0} \frac{\sec^2 x}{2} = 2$$

6-23

Prove by induction that, for any positive integer n, $\lim_{x\to\infty} x^n e^{-x} = 0$.

****

$$\lim_{x\to\infty} x^n e^{-x} = \lim_{x\to\infty} \frac{x^n}{e^x}$$

Show true for n = 1 : $\lim_{x\to\infty} \frac{x}{e^x}$ indeterminate form $\frac{\infty}{\infty}$

By l'hopital's rule, $= \lim_{x\to\infty} \frac{1}{e^x} = 0$. True for n = 1.

Assume true for n = k. Show true for n = k+1:

$\lim_{n\to\infty} \frac{x^k}{e^x} = 0$ is given (assumed). Must show $\lim_{x\to\infty} \frac{x^{k+1}}{e^x} = 0$.

$\lim_{x\to\infty} \frac{x^{k+1}}{e^x}$ indeterminate form $\frac{\infty}{\infty}$; by l'hopital's rule, $= \lim_{x\to\infty} \frac{(k+1)x^k}{e^x} = (k+1) \lim_{x\to\infty} \frac{x^k}{e^x} = (k+1)(0) = 0$.

True for n = k+1.

Therefore, $\lim_{x\to\infty} x^n e^{-x} = 0$ for all positive integers n.

IMPROPER INTEGRALS

6-24

Determine whether the improper integral is convergent or divergent. If it is convergent, evaluate it.

$$\int_{-\infty}^{2} \frac{dx}{x^2 + 4}$$

****

Because the integral is improper,

$$\int_{-\infty}^{2} \frac{dx}{x^2+4} = \lim_{a \to -\infty} \int_{a}^{2} \frac{dx}{x^2+4} = \lim_{a \to -\infty} \frac{1}{2} \tan^{-1}\left(\frac{x}{2}\right)\Bigg|_{a}^{2}$$

$$= \lim_{a \to -\infty} \left[\tfrac{1}{2}\tan^{-1}(1) - \tfrac{1}{2}\tan^{-1}\left(\tfrac{a}{2}\right)\right]$$

$$= \tfrac{1}{2}\tan^{-1}(1) - \tfrac{1}{2}\lim_{a \to -\infty}\left[\tan^{-1}\left(\tfrac{a}{2}\right)\right]$$

$$= \tfrac{1}{2}\cdot\frac{\pi}{4} - \tfrac{1}{2}\left(-\tfrac{\pi}{2}\right) = \tfrac{\pi}{8} + \tfrac{\pi}{4} = \frac{3\pi}{8}$$

6-25

Evaluate $\int_0^\infty e^{-4x}\,dx$

****

$$\int_0^\infty e^{-4x}\,dx = \lim_{a \to \infty} \int_0^a e^{-4x}\,dx$$

$$= \lim_{a \to \infty}\left[-\frac{1}{4}e^{-4x}\right]_0^a$$

$$= \lim_{a \to \infty}\left(-\frac{1}{4}e^{-4a} + \frac{1}{4}\right)$$

$$= \frac{1}{4}$$

6-26

Evaluate the integral:

$$\int_0^5 \frac{1}{\sqrt{x}} + \frac{1}{\sqrt{5-x}} \, dx$$

****

$$\int_0^5 \frac{1}{\sqrt{x}} + \frac{1}{\sqrt{5-x}} \, dx = \lim_{s \to 0^+} \int_s^1 \frac{1}{\sqrt{x}} \, dx$$

$$+ \lim_{t \to 5^-} \int_1^t \frac{1}{\sqrt{5-x}} \, dx$$

$$= \lim_{s \to 0^+} 2\sqrt{x} \Big|_s^1 + \lim_{t \to 5^-} -2\sqrt{5-x} \Big|_1^t$$

$$= \lim_{s \to 0^+} 2(1-\sqrt{s}) + \lim_{t \to 5^-} -2(\sqrt{5-t} - 2)$$

$$= 2 + 4 = 6$$

6-27

Evaluate $\int_3^\infty \frac{1}{x^{3/2}} dx$.

****

$$\int_3^\infty \frac{1}{x^{3/2}} dx = \lim_{t \to \infty} \int_3^t x^{-3/2} dx = \lim_{t \to \infty} -2x^{-1/2} \Big]_3^t$$

$$= \lim_{t \to \infty} \frac{-2}{\sqrt{t}} - \left(\frac{-2}{\sqrt{3}}\right) = \frac{2}{\sqrt{3}}.$$

6-28

Evaluate the integral $\int_0^1 \frac{1}{(x-1)^2}\, dx$.

****

THE INTEGRAL IS IMPROPER BECAUSE $\frac{1}{(x-1)^2}$ IS UNDEFINED AT 1, THE UPPER LIMIT OF INTEGRATION.

$$\int_0^1 \frac{1}{(x-1)^2}\,dx = \lim_{b\to 1^-} \int_0^b (x-1)^{-2}\,dx$$

$$= \lim_{b\to 1^-} \left(-(x-1)^{-1}\Big|_0^b\right)$$

$$= \lim_{b\to 1^-} \left(\frac{1}{1-x}\Big|_0^b\right)$$

$$= \lim_{b\to 1^-} \left(\frac{1}{1-b} - 1\right)$$

$$= \infty.$$

THE INTEGRAL DIVERGES TO INFINITY.

6-29

$\int_{-1}^3 1/x\, dx$ is (a) 8/9 (b) ln 3 (c) -1 + ln 3 (d) -10/9 (e) a divergent improper integral.

****

$f(x) = \frac{1}{x}$ has a vertical asymptote at $x=0$ so $\int_{-1}^3 \frac{1}{x}\,dx = \int_{-1}^0 \frac{1}{x}\,dx + \int_0^3 \frac{1}{x}\,dx$ provided each of these 2 are convergent.

But $\int_0^3 \frac{1}{x}\,dx = \lim_{t\to 0^+} \int_t^3 \frac{1}{x}\,dx = \lim_{t\to 0^+} \ln x\Big|_t^3 = \lim_{t\to 0^+}(\ln 3 - \ln t)$ $= +\infty$, thus $\int_{-1}^3 \frac{1}{x}\,dx$ is divergent.

6-30

Evaluate $\int_0^2 \frac{dx}{(x-1)^2}$ to determine if it is convergent.

****

We note that if $x = 1$, $\frac{1}{(x-1)^2}$ becomes undefined. We write

$$\int_0^2 \frac{dx}{(x-1)^2} = \int_0^1 \frac{dx}{(x-1)^2} + \int_1^2 \frac{dx}{(x-1)^2}$$

In integrating $\int_0^1 \frac{dx}{(x-1)^2}$ we write

$$\int_0^t \frac{dx}{(x-1)^2} = \frac{-1}{x-1}\Bigg/_0^t$$

$$= \frac{-1}{-(1-x)}\Bigg/_0^t$$

$$= \left(\frac{1}{1-t}\right) - \left(\frac{1}{1-0}\right)$$

$$= \frac{1}{1-t} - 1 \Longrightarrow \infty \text{ as } t \uparrow 1$$

Hence, $\int_0^1 \frac{dx}{(x-1)^2}$ diverges,

i.e. $\int_0^2 \frac{dx}{(x-1)^2}$, of which $\int_0^1 \frac{dx}{(x-1)^2}$ is a part, also diverges.

6-31

Classify the following integrals as convergent vs. divergent and evaluate if convergent: (a) $\int_3^\infty 1/x^2\,dx$ (b) $\int_{-1}^3 1/x^2\,dx$.

****

(a) $\int_3^\infty \frac{1}{x^2}\,dx = \lim_{t\to\infty} \int_3^t \frac{1}{x^2}\,dx = \lim_{t\to\infty} -\frac{1}{x}\Big|_3^t = \lim_{t\to\infty}\left(-\frac{1}{t}+\frac{1}{3}\right)$

$= \frac{1}{3}$, so it is convergent.

(b) $\int_{-1}^3 \frac{1}{x^2}\,dx = \int_{-1}^0 \frac{1}{x^2}\,dx + \int_0^3 \frac{1}{x^2}\,dx$, provided these two are convergent. $\int_0^3 \frac{1}{x^2}\,dx = \lim_{t\to 0^+} \int_t^3 \frac{1}{x^2}\,dx =$

$\lim_{t\to 0^+} -\frac{1}{x}\Big|_t^3 = \lim_{t\to 0^+}\left(-\frac{1}{3}+\frac{1}{t}\right) = +\infty$, therefore

$\int_{-1}^3 \frac{1}{x^2}\,dx$ is divergent.

6-32

Evaluate $\displaystyle\int_0^\infty \frac{x}{(x^2+5)^2}\,dx$

****

Letting $u = x^2 + 5$,

$$\int \frac{x\,dx}{(x^2+5)^2} = \frac{1}{2}\int \frac{du}{u^2} = \frac{-1}{2u} = \frac{-1}{2(x^2+5)}$$

$$\int_0^\infty \frac{x\,dx}{(x^2+5)^2} = \lim_{N\to\infty} \int_0^N \frac{x\,dx}{(x^2+5)^2}$$

$$= \lim_{N\to\infty} \left(\frac{-1}{2(x^2+5)}\right)\Bigg|_0^N = \lim_{N\to\infty}\left(\frac{-1}{2(N^2+5)} + \frac{1}{10}\right) = \frac{1}{10}$$

6-33

Evaluate $\int_0^1 \ln(x)\, dx$

****

$$\int \ln x\, dx = x \ln x - \int dx = x \ln x - x + C$$

$u = \ln x \qquad dv = dx$

$du = \frac{1}{x}\, dx \qquad v = x$

$$\int_0^1 \ln x\, dx = \lim_{a \to 0^+} \int_a^1 \ln x\, dx = \lim_{a \to 0^+} (x \ln x - x)\Big|_a^1$$

$$= (1 \ln 1 - 1) - \lim_{a \to 0^+} (a \ln a - a)$$

$$\left[\lim_{a \to 0^+} a \ln a = \lim_{a \to 0^+} \frac{\ln a}{1/a} = \lim_{a \to 0^+} \frac{1/a}{-1/a^2} = \lim_{a \to 0^+} (-a) = 0 \right]$$

$$= (0 - 1) - (0 - 0) = \underline{\underline{-1}}$$

6-34

Evaluate the following definite integral.

$$\int_0^4 \frac{dx}{|x - 1|^{1/2}}$$

****

Since $x - 1 = 0$ at $x = 1$, $x - 1 < 0$ on $[0, 1)$ and $x - 1 > 0$ at $(1, 4]$;

$$\int_0^4 \frac{dx}{|x-1|^{1/2}} = \int_0^1 \frac{dx}{(1-x)^{1/2}} + \int_1^4 \frac{dx}{(x-1)^{1/2}}$$

$$= \lim_{a \to 1^-} \int_0^a (1-x)^{-1/2} dx + \lim_{b \to 1^+} \int_b^4 (x-1)^{-1/2} dx$$

$$= \lim_{a\to 1^-} \left(-2(1-x)^{1/2}\Big|_0^a\right) + \lim_{b\to 1^+} \left(2(x-1)^{1/2}\Big|_b^4\right)$$

$$= \lim_{a\to 1} \left[-2(1-a)^{1/2} + 2\right] + \lim_{b\to 1^+} \left[2\sqrt{3} - 2(b-1)^{1/2}\right]$$

$$= \boxed{2 + 2\sqrt{3}}$$

6-35

Show that the surface obtained by rotating $y = \frac{1}{x}$ $(0 < x \leq 1)$ about the y -axis has infinite surface area, but encloses a finite volume.

****

$$S.A. = \int_0^1 2\pi x \sqrt{1 + \left(\frac{-1}{x^2}\right)^2}\, dx$$

$$= \lim_{a\to 0+} \int_a^1 2\pi x \sqrt{1 + \frac{1}{x^4}}\, dx$$

$$> \lim_{a\to 0+} \int_a^1 2\pi x \sqrt{\frac{1}{x^4}}\, dx$$

$$= \lim_{a\to 0+} \int_a^1 2\pi \cdot \frac{1}{x}\, dx = +\infty.$$

$$V = \int_1^\infty \frac{\pi}{y^2}\, dy = \lim_{b\to\infty} \int_1^b \frac{\pi}{y^2}\, dy$$

$$= \lim_{b\to\infty} -\frac{\pi}{y}\Big|_1^b = \pi.$$

6-36

State whether to following improper integral is convergent or divergent. For convergent, state the value.

$$\int_1^3 \frac{2\ dx}{(x-2)^{4/3}}$$

****

Since the denominator goes to zero at $x=2$ we must rewrite:

$$\int_1^3 \frac{2\,dx}{(x-2)^{4/3}} = \int_1^2 \frac{2\,dx}{(x-2)^{4/3}} + \int_2^3 \frac{2\,dx}{(x-2)^{4/3}}$$

$$= \lim_{a\to 2^-} \int_1^a \frac{2\,dx}{(x-2)^{4/3}} + \lim_{b\to 2^+} \int_b^3 \frac{2\,dx}{(x-2)^{4/3}}$$

$$= \lim_{a\to 2^-} \left.\frac{2(x-2)^{-1/3}}{-1/3}\right]_1^a + \lim_{b\to 2^+} \left.\frac{2(x-2)^{-1/3}}{-1/3}\right]_b^3$$

$$= -6\lim_{a\to 2^-}\left[(a-2)^{-1/3} - (+1-2)^{-1/3}\right] - 6\lim_{b\to 2^+}\left[(3-2)^{-1/3} - (b-2)^{-1/3}\right]$$

$$= -6(-\infty + 1 + 1 - \infty) \to +\infty \quad \therefore \text{divergent}$$

6-37

Evaluate the following: $\int_1^{\infty} x^{-5/4}\,dx$

****

$$\int_1^{\infty} x^{-5/4}\,dx = \lim_{b\to\infty} \int_1^b x^{-5/4}\,dx = \lim_{b\to\infty} \left[\frac{x^{-1/4}}{-1/4}\right]_1^b$$

$$= \lim_{b\to\infty} \left[\frac{b^{-1/4}}{-1/4} - \frac{1^{-1/4}}{-1/4}\right] = 0 - \frac{1}{-1/4} = 4$$

6-38

Sketch $f(x) = \dfrac{1 + \ln x}{x}$ and determine the area enclosed between the curve and the x axis, as x approaches ∞.

****

$f(x) = \dfrac{1+\ln x}{x}$

Domain of $f(x)$ $(0,\infty)$

X intercept when $f(x)=0 \rightarrow 1+\ln x=0 \rightarrow \ln x=-1 \rightarrow x=e^{-1}=1/e$.

Vertical asymptote at $x=0$. $\lim\limits_{x\to 0^+} \dfrac{1+\ln x}{x} = \dfrac{1-\infty}{0^+} = -\infty$

Horizontal asymptote $\lim\limits_{x\to\infty} \dfrac{1+\ln x}{x} = \lim\limits_{x\to\infty} \dfrac{1/x}{1} = 1/\infty = 0$ (using l'hopitals rule)

$f'(x) = \dfrac{\left(\frac{1}{x}\right)x-(1+\ln x)}{x^2} = \dfrac{-\ln x}{x^2}$

$-\ln x$		+ + + +	− − − − − −
x^2		+ + + +	+ + + + + +
$f'(x)$	0	+ ↗	1 − ↘

Horizontal tangent at $x=1$, $y=1$

Increasing $(0,1]$

Decreasing $[1,\infty)$

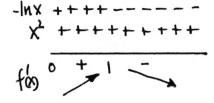

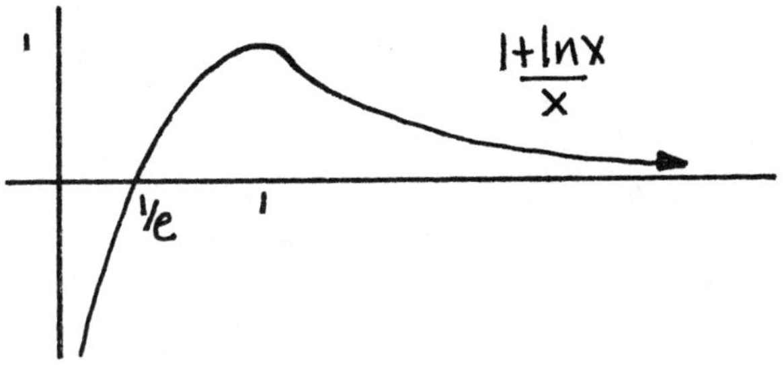

Is the area finite?

$A = \displaystyle\int_{1/e}^{\infty} \frac{1+\ln x}{x}\,dx = \lim_{b\to\infty} \int_{1/e}^{b} \frac{1+\ln x}{x}\,dx$

let $u = 1+\ln x$

$\dfrac{du}{dx} = \dfrac{1}{x} \rightarrow dx = x\,du$

$\int u\,du = \dfrac{u^2}{2}+C$. So $A = \lim\limits_{b\to\infty} \left.\dfrac{(1+\ln x)^2}{2}\right|_{1/e}^{b} = \dfrac{(1+\ln b)^2}{2} - \dfrac{(1+\ln 1/e)^2}{2}$

But $\ln b \to \infty$ as $b \to \infty$.

Thus the area between $\dfrac{1+\ln x}{x}$ and the x axis is infinite

6-39

Evaluate the following improper integral: $\int_0^{\infty} \frac{8}{x^2+4}\,dx$.

****

By definition $\int_0^{\infty} \frac{8}{x^2+4}\,dx = \lim_{t\to\infty} \int_0^{t} \frac{8}{x^2+4}\,dx$.

So $\int_0^{\infty} \frac{8}{x^2+4}\,dx = \lim_{t\to\infty} \int_0^{t} \frac{8}{x^2+4}\,dx$

$$= 8 \lim_{t\to\infty} \int_0^{t} \frac{1}{x^2+4}\,dx$$

$$= 8 \lim_{t\to\infty} \int_0^{t} \frac{\frac{1}{4}}{\left(\frac{x}{2}\right)^2+1}\,dx$$

$$= 2(2) \lim_{t\to\infty} \int_0^{t} \frac{\frac{1}{2}}{\left(\frac{x}{2}\right)^2+1}\,dx$$

$$= 4 \lim_{t\to\infty} \left(\arctan \frac{x}{2}\right)\Big|_0^{t}$$

$$= 4 \lim_{t\to\infty} \left(\arctan \frac{t}{2}\right) - \lim_{t\to\infty} (\arctan 0)$$

$$= 4\left(\frac{\pi}{2} - 0\right)$$

$$= 2\pi$$

7
TAYLOR POLYNOMIALS AND APPROXIMATION

APPROXIMATIONS WITH TAYLOR POLYNOMIALS

7-1

Find the fourth degree Taylor's Polynomial of the function $f(x) = xe^x$ at the number a=0.

****

$f(x) = xe^x$ $\qquad f(0) = 0$

$f'(x) = xe^x + e^x$ $\qquad f'(0) = 1$

$f''(x) = xe^x + e^x + e^x = xe^x + 2e^x$ $\qquad f''(0) = 2$

$f'''(x) = xe^x + 3e^x$ $\qquad f'''(0) = 3$

$f^{(4)}(x) = xe^x + 4e^x$ $\qquad f^{(4)}(0) = 4$

$$\therefore\ f(x) = xe^x \approx 0 + \frac{1\,(x-0)}{1!} + \frac{2\,(x-0)^2}{2!} + \frac{3\,(x-0)^3}{3!} + \frac{4\,(x-0)^4}{4!}$$

$$= x + x^2 + \frac{x^3}{2} + \frac{x^4}{6}$$

7-2

State Taylor's formula for nth-degree polynomials and Lagrange form of the remainder in order to approximate the value of f(x). Find the third-degree Taylor polynomial of the cosine function at $\frac{1}{3}\pi$ and the Lagrange form of the remainder.

****

Taylor's formula for nth-degree polynomial is given as

$$P_n(x) = f(a) + \left[\frac{f'(a)}{1!}(x-a)\right] + \left[\frac{f''(a)}{2!}(x-a)^2\right] + \dots\dots + \frac{f^n(a)}{n!}(x-a)^n \quad \dots\dots(1)$$

and Lagrange form of the remainder is given as

$$R_n(x) = \frac{f^{(n+1)}(c)}{n+1}(x-a)^{n+1} \quad \dots\dots(2)$$

where c is between a and x.

Hence, $f(x) = P_n(x) + R_n(x)$

For the given function, let $f(x) = \cos x$. We then have, using Taylor's formula

$$P_3(x) = f\left(\frac{\pi}{3}\right) + \left[\frac{f'\left(\frac{\pi}{3}\right)}{1!}\left(x-\frac{\pi}{3}\right)\right] + \left[\frac{f''\left(\frac{\pi}{3}\right)}{2!}\left(x-\frac{\pi}{3}\right)^2\right] + \left[\frac{f'''\left(\frac{\pi}{3}\right)}{3!}\left(x-\frac{\pi}{3}\right)^3\right] \quad \dots\dots(3)$$

Now, $f(x) = \cos x$; hence, $f\left(\frac{\pi}{3}\right) = 0{\cdot}5$

$f'(x) = -\sin x$; $\quad f'\left(\frac{\pi}{3}\right) = -0{\cdot}866$

$f''(x) = -\cos x$; $\quad f''\left(\frac{\pi}{3}\right) = -0{\cdot}5$

$f'''(x) = \sin x$; $\quad f'''\left(\frac{\pi}{3}\right) = 0{\cdot}866$

Substituting these values in Eq.(3) we get

$$P_3(x) = 0{\cdot}5 + \left[\frac{-0{\cdot}866}{1!}\left(x-\frac{\pi}{3}\right)\right] + \left[\frac{-0{\cdot}5}{2!}\left(x-\frac{\pi}{3}\right)^2\right] + \left[\frac{0{\cdot}866}{3!}\left(x-\frac{\pi}{3}\right)^3\right]$$

$$= 0{\cdot}500 + \left[0{\cdot}866\left(x-\frac{\pi}{3}\right)\right] - \left[0{\cdot}250\left(x-\frac{\pi}{3}\right)^2\right] + \left[0{\cdot}140\left(x-\frac{\pi}{3}\right)^3\right] \quad \underline{\text{Ans.}}$$

Similarly, Lagrange form of the remainder gives us:

$$R_3(x) = \frac{f^{(3+1)}(c)}{(3+1)!}\left(x - \frac{\pi}{3}\right)^{(3+1)}$$

$$= \frac{f^4(c)}{4!}\left(x - \frac{\pi}{3}\right)^4$$

$$= \frac{1}{24}(\cos c)\left(x - \frac{\pi}{3}\right)^4$$

where c is between $\frac{\pi}{3}$ and x.

We note that as $|\cos c| \leq 1$, we can write for all x,

$$R_3(x) \leq \frac{1}{24}(\cos c)\left(x - \frac{\pi}{3}\right)^4 \quad \underline{\text{Ans.}}$$

7-3

A particle moves along the numbers line. At time $t = 1$ its position, velocity and acceleration are all observed to be 1. Use an appropriate Taylor polynomial to approximate its position at time $t = 2$.

****

Let the position function of the particle be $s(t)$.

Expanding $s(t)$ in a Taylor polynomial around $t=1$ gives

$$s(t) \approx s(1) + s'(1)(t-1) + \frac{s''(1)}{2}(t-1)^2$$

Since $s'(t) = v(t)$, we have $s'(1) = 1$

and $s''(t) = a(t)$, we have $s''(1) = 1$

thus: $s(t) \approx 1 + 1\cdot(t-1) + \frac{1}{2}(t-1)^2$

in particular, for $t = 2$,

$$s(2) = 1 + 1 + \frac{1}{2} = 2\tfrac{1}{2}$$

7-4

Find the Taylor polynomial $T_3(x)$ for the function $f(x) = \frac{5x}{2+4x}$ at the point $x_0 = 0$.

****

$$T_3(x) = \frac{f'''(x_0)}{3!}(x-x_0)^3 + \frac{f''(x_0)}{2!}(x-x_0)^2 + \frac{f'(x_0)}{1!}(x-x_0) + f(x_0)$$

$$f(x) = \frac{5x}{4+2x} \quad \text{so} \quad f(0) = 0$$

$$f'(x) = \frac{5(4+2x)-10x}{(4+2x)^2} = \frac{20}{(4+2x)^2} \quad \text{so} \quad f'(0) = 5/4$$

$$f''(x) = \frac{-80}{(4+2x)^3} \quad \text{so} \quad f''(0) = -5/4$$

$$f'''(x) = \frac{240}{(4+2x)^4} \quad \text{so} \quad f'''(0) = 15/8.$$

So

$$T_3(x) = \frac{15}{8\cdot 6}x^3 - \frac{5}{4\cdot 4}x^2 + \frac{5}{4\cdot 1}x$$

$$= \frac{15}{48}x^3 - \frac{5}{16}x^2 + \frac{5}{4}x$$

7-5

Use a Taylor polynomial of degree 5 to approximate the function f(x)= sin x.

****

The Taylor polynomial of degree 5 for $f(x) = \sin x$ is given by

$$f(x) = f(0) + \frac{f'(0)}{1!}x + \frac{f''(0)}{2!}x^2 + \frac{f'''(0)}{3!}x^3 + \frac{f^{iv}(0)}{4!} + \frac{f^{v}(0)}{5!}x^5$$

$f(x) = \sin x$ so $f(0) = 0$

$f'(x) = \cos x$ so $f'(0) = 1$

$f''(x) = -\sin x$ so $f''(0) = 0$

$f'''(x) = -\cos x$ so $f'''(0) = -1$

$f^{iv}(x) = \sin x$ so $f^{iv}(0) = 0$

$f^{v}(x) = \cos x$ so $f^{v}(0) = 1$

Therefore $f(x) = \sin x$ is <u>approximated</u> by a 5th degree Taylor polynomial as follows:

$$\sin x \approx 0 + \frac{1}{1!}x + 0 + \frac{(-1)}{3!}x^3 + 0 + \frac{1}{5!}x^5$$

$$= x - \frac{1}{6}x^3 + \frac{1}{120}x^5$$

7-6

Write the fourth degree Taylor polynomial centered about the origin for the function $f(x) = e^{-2x}$.

****

$$f'(x) = -2e^{-2x}, \quad f'(0) = -2$$

$$f''(x) = 4e^{-2x}, \quad f''(0) = 4$$

$$f'''(x) = -8e^{-2x}, \quad f'''(0) = -8$$

$$f^{(4)}(x) = 16e^{-2x}, \quad f^{(4)}(0) = 16$$

$P_4(x)$ = 4th degree Taylor polynomial for $f(x)$, centered at $x=0$

$$= 1 - 2x + \frac{4}{2!}x^2 - \frac{8}{3!}x^3 + \frac{16}{4!}x^4$$

$$= 1 - 2x + 2x^2 - \frac{4}{3}x^3 + \frac{2}{3}x^4$$

$$\left[P_4(x) = f(0) + \frac{f'(0)}{1!}x + \frac{f''(0)}{2!}x^2 + \frac{f'''(0)}{3!}x^3 + \frac{f^{(4)}(0)}{4!}x^4 \right]$$

7-7

Approximate the following integral using Taylor Polynomials:

$$\int_0^{1/2} e^{-3x^2}\,dx \quad .$$

****

$$e^x = 1 + x + \frac{x^2}{2!} + \frac{x^3}{3!} + \frac{x^4}{4!} + \cdots\cdots \quad (\text{for } a=0)$$

Replace x by $-3x^2$:

$$e^{-3x^2} = 1 - 3x^2 + \frac{9x^4}{2!} - \frac{27x^6}{3!} + \frac{81x^8}{4!} - \cdots\cdots$$

$$\int_0^{\frac{1}{2}} e^{-3x^2}dx = \int_0^{\frac{1}{2}} \left(1 - 3x^2 + \frac{9x^4}{2!} - \frac{27x^6}{3!} + \frac{81x^8}{4!} - \cdots\right)dx$$

$$= x - x^3 + \frac{9}{2}\frac{x^5}{5} - \frac{27}{6}\frac{x^7}{7} + \frac{81}{24}\frac{x^9}{9} - \cdots \Big|_0^{\frac{1}{2}}$$

$$= \frac{1}{2} - \frac{1}{8} + \frac{9}{10\cdot 32} - \frac{27}{42\cdot 128} + \frac{81}{216\cdot 512} - \cdots - 0$$

$$= 0.5000 - 0.1250 + 0.0281 - 0.0050 + 0.0007 - \cdots$$

$$= 0.3988 - \cdots$$

Therefore, $\int_0^{\frac{1}{2}} e^{-3x^2}dx = 0.399$

7-8

Find the Taylor polynomial, $P_3(x)$ for $f(x) = xe^x$, i.e., find the first four terms of the Taylor Series expansion of $f(x) = xe^x$.

****

$f(0) = 0$

$f'(x) = xe^x + e^x; \; f'(0) = 1$

$f''(x) = xe^x + 2e^x; \; f''(0) = 2$

$f'''(x) = xe^x + 3e^x; \; f'''(0) = 3$

$$P_3(x) = 0 + x + \frac{2}{2!}x^2 + \frac{3}{3!}x^3$$

$$P_3(x) = 0 + x + x^2 + \frac{1}{2}x^3$$

ESTIMATES ON THE REMAINDER TERM

7-9

Find the second degree Taylor polynomial for $f(x) = \sqrt{x}$, centered about $a = 100$. Also obtain a bound for the error in using this polynomial to approximate $\sqrt{100.1}$.

****

$$f'(x) = \frac{1}{2\sqrt{x}}, \quad f''(x) = -\frac{1}{4x^{3/2}}, \quad f'''(x) = \frac{3}{8x^{5/2}}$$

$P_2(x)$ = 2nd degree Taylor polynomial for $f(x)$, centered about $a = 100$

$$= f(100) + f'(100)(x-100) + \frac{f''(100)}{2!}(x-100)^2$$

$$= 10 + \frac{1}{20}(x-100) - \frac{1}{8000}(x-100)^2$$

$$|\text{error}| \le \left(\max_{100 \le x \le 100.1} f'''(x)\right)\left(100.1-100\right)^3 / 3!$$

$$= \frac{3}{800000}(0.001)/6$$

$$= \underline{6 \times 10^{-10}}$$

7-10

Find an approximation for sin(.1) accurate to 6 decimal places (The .1 is in radians).

****

$$\sin x = x - \frac{x^3}{3!} + \frac{x^5}{5!} - + \cdots + \frac{\sin^{[n]}(0)}{n!}x^n + R_n(x),$$

$$R_n(x) = \frac{\sin^{[n+1]}(\xi)}{(n+1)!}x^{n+1}, \quad 0 < \xi < x.$$

since $\sin^{[4]}(x) = \sin x$, $\sin^{[4]}(0) = 0$ and

$$R_4(x) = \frac{\cos(\xi)}{5!}x^5.$$

$$|R_4(.1)| = \frac{|\cos(\xi)|}{5!}(.1)^5 < \frac{.00001}{120}$$

$$< .0000001, \text{ so}$$

$$\sin(.1) = .1 - \frac{.001}{6} \pm 10^{-7},$$

$$\sin(.1) = .099833.$$

(A calculator gives $\sin(.1) = .0998334$)

7-11

a) Give the 4th degree Taylor polynomial for $f(x) = \sqrt{x}$ about the point $x = 4$.
b) Using this polynomial, approximate $\sqrt{4.2}$.
c) Give the maximum error for this approximation.

****

a) $f(x) = f(c) + f'(c)(x-c) + \frac{f''(c)}{2!}(x-c)^2 + f'''(c)(x-c)^3 + f^{iv}(c)(x-c)^4$

$+ R_4$

k	$f^k(x)$	$f^k(4)$
0	$x^{1/2}$	2
1	$\frac{1}{2}x^{-1/2}$	$\frac{1}{4}$
2	$-\frac{1}{4}x^{-3/2}$	$-\frac{1}{32}$
3	$\frac{3}{8}x^{-5/2}$	$\frac{3}{256}$
4	$-\frac{15}{16}x^{-7/2}$	$\frac{-15}{(16)(128)}$
5	$\frac{105}{32}x^{-9/2}$	$\frac{105}{(32)(512)}$

$$\sqrt{x} = 2 + \frac{1}{4}(x-4) + \frac{1}{2}\left(\frac{-1}{32}\right)(x-4)^2$$

$$+\left(\frac{1}{6}\right)\left(\frac{3}{256}\right)(x-4)^3 + \frac{1}{24}\left(\frac{-15}{16\cdot 128}\right)(x-4)^4 + R_4$$

$$= \boxed{2 + \frac{x-4}{4} + \frac{(x-4)^2}{64} + \frac{(x-4)^3}{512} - \frac{5(x-4)^4}{16384} + R_4}$$

b) $\sqrt{4.2} \doteq 2 + \frac{.2}{4} + \frac{.04}{64} + \frac{.008}{512} - \frac{.008}{16384}$

$= \boxed{2.049390137}$

c) Since the series is alternating the error is given by the next term.

$$R_4 = \frac{1}{5!}\left(\frac{105}{32}\right)(z)^{-9/2}(4.2)^5$$

$$= \frac{1}{5!}\left(\frac{105}{32}\right)\frac{1}{(4.2)^{9/2}}(4.2)^5 \doteq \boxed{.0000000171}$$

7-12

(a) Use the 3rd degree Taylor Polynomial of $f(x) = \sqrt{x}$ about $x=4$ to approximate $\sqrt{6}$.

(b) Use the remainder term to give an upper bound for the error in the above approximation.

****

$f(x) = x^{1/2}$, $f'(x) = \frac{1}{2}x^{-1/2}$, $f''(x) = \frac{-1}{4}x^{-3/2}$, $f'''(x) = \frac{3}{8}x^{-5/2}$

$f(4) = 2$, $f'(4) = \frac{1}{4}$, $f''(4) = \frac{-1}{32}$, $f'''(4) = \frac{3}{256}$

$$P_3(x) = 2 + \frac{1}{4}(x-4) + \frac{\left(\frac{-1}{32}\right)}{2}(x-4)^2 + \frac{\left(\frac{3}{256}\right)}{6}(x-4)^3$$

$$= 2 + \frac{1}{4}(x-4) - \frac{1}{64}(x-4)^2 + \frac{1}{512}(x-4)^3$$

(a) $$\sqrt{6} = f(6) \approx P_3(6) = 2 + \frac{2}{4} - \frac{2^2}{64} + \frac{2^3}{512}$$

$$= 2 + \frac{1}{2} - \frac{1}{16} + \frac{1}{64} = 2.453125$$

(b) $f^{(4)}(x) = \frac{-15}{16}x^{-7/2}$, and for some $4 \le c \le 6$,

$$|\text{error}| = \left| \frac{\left(\frac{-15}{16}\right)c^{-7/2}}{24} \cdot 2^4 \right| \le \frac{15}{24} \cdot 4^{-7/2}$$

$$\le \frac{5}{1024} \le 0.005$$

7-13

a) Find the third degree Taylor polynomial, with remainder, centered at $x = 1$ for $f(x) = \ell n\, x$.

b) Use the result from part (a) to approximate $\ell n(1.2)$ and estimate the accuracy of this approximation.

****

a) If $f(x) = \ln x$ then $f(1) = 0$

and: $f'(x) = x^{-1}$ $\quad f'(1) = 1$

$f''(x) = -x^{-2}$ $\quad f''(1) = -1$

$f'''(x) = 2x^{-3}$ $\quad f'''(1) = 2$

So the required polynomial is

$$P(x) = 0 + \frac{1}{1!}(x-1) - \frac{1}{2!}(x-1)^2 + \frac{2}{3!}(x-1)^3$$

Since $f^{(IV)}(x) = -6x^{-4}$ the remainder term is

$$R_3(x) = \frac{-6c^{-4}}{4!}(x-1)^4 = \frac{-(x-1)^4}{4c^4}$$

(where c is some number between 1 and x)

b) Evaluating $P(1.2)$ gives

$$\ln(1.2) \approx 0.2 - \tfrac{1}{2}(0.2)^2 + \tfrac{1}{3}(0.2)^3 = 0.18266$$

Now $$|R_3(1.2)| = \left|\frac{-(0.2)^4}{4c^4}\right| < \frac{.0016}{4} = .0004$$

Since $.0004 < .0005 = .5 \times 10^{-3}$ we know that the approximation is accurate to 3 decimal places.

NEWTON'S METHOD OF SOLVING EQUATIONS

7-14

Find one root for the equation $f(x) = x^3 - x - 7$ accurate to six decimal places.

****

X	f(X)
2	-1
2.1	.161

root →

$$X_{n+1} = X_n - \frac{X_n^3 - X_n - 7}{3X_n^2 - 1}$$

$$= \frac{2X_n^3 + 7}{3X_n^2 - 1}$$

$X_0 = 2$

$X_1 = 2.0909091$

$X_2 = 2.0867543$

$X_3 = 2.0867453$

$X_4 = 2.0867453$

ANS: $X = 2.086745$

7-15

Use Newton's Method to get three successive approximations to the root of $x^3 + x^2 - 1 = 0$ which lies between 0 and 1.

****

Different answers will be correct, depending on the first approximate value chosen, x_0. The next two approximations will be $x_1 = x_0 - \frac{f(x_0)}{f'(x_0)}$ and $x_2 = x_1 - \frac{f(x_1)}{f'(x_1)}$, where $f(x) = x^3 + x^2 - 1$.

Note: $f'(x) = 3x^2 + 2x$.

If you choose $x_0 = 1$, then $x_1 = 1 - \frac{1}{5} = \frac{4}{5}$, and

$$x_2 = \frac{4}{5} - \frac{\left(\frac{4}{5}\right)^3 + \left(\frac{4}{5}\right)^2 - 1}{3\left(\frac{4}{5}\right)^2 + 2\left(\frac{4}{5}\right)} = \frac{4}{5} - \frac{19/125}{88/25} = \frac{4}{5} - \frac{19}{440} = \frac{333}{440}.$$

7-16

Use Newton's method with starting value $x_0 = -1$ to approximate the solution of $x^3 - 5 = 0$. Apply the method twice.

****

LET $f(x) = x^3 - 5$

THEN $f'(x) = 3x^2$.

$$x_{n+1} = x_n - \frac{f(x_n)}{f'(x_n)}$$

$$x_0 = -1$$

$$x_1 = -1 - \frac{(-1)^3 - 5}{3(-1)^2} = -1 - \frac{-6}{3} = 1$$

$$x_2 = 1 - \frac{1^3 - 5}{3(1)^2} = 1 - \frac{-4}{3} = \frac{7}{3}$$

7-17

Suppose $f(x)$ is a function such that $f'(x) > 0$ and $f''(x) > 0$ (everywhere). Suppose r is a root of $f(x)$, i.e. $f(r) = 0$. Suppose r is to be approximated using Newton's Method. Let r_1 be the first approximation and r_2 the next approximation. If $r_1 < r$, what can you conclude about r_2:

$r_2 < r_1 < r$ $\quad r_1 < r_2 < r$ $\quad r_1 < r < r_2$ $\quad$ none of the preceding

Give reasons for your answer (a suitable picture will suffice)

****

Since $f'(x) > 0$ and $f''(x) > 0$, the function $f(x)$ is everywhere increasing and concave up (see sketch below)

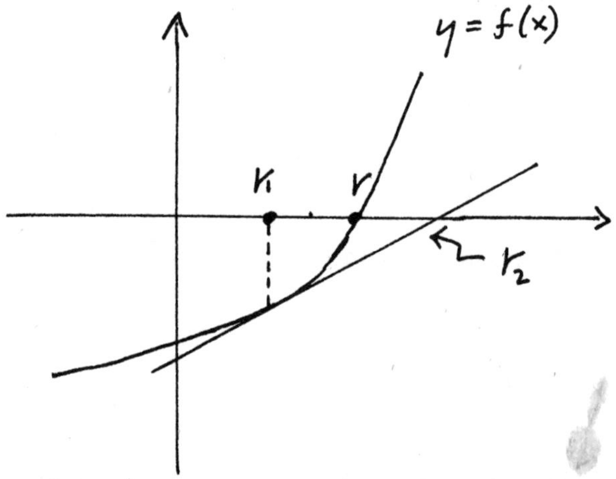

In Newton's Method, r_2 is obtained from r_1 by drawing the tangent line to the curve $y = f(x)$ at $(r_1, f(r_1))$ and seeing where it intersects the x-axis.

The picture makes it clear that under the hypotheses of the problem,

$$r_1 < r < r_2$$

8 SEQUENCES AND SERIES

SEQUENCES

8-1

If $\dfrac{3n - 1}{n + 1} < x_n < \dfrac{3n^2 + 6n + 2}{n^2 + 2n + 1}$ for all positive integers n, then
(a) $\lim_{n\to\infty} x_n = L$ where $-1 \leq L \leq 2$ (b) $\lim_{n\to\infty} x_n = 3$ (c) $\{x_n\}$ is bounded but may be divergent (d) $\{x_n\}$ is monotonic (e) $\{x_n\}$ must be divergent.

****

$\lim_{n\to\infty} \dfrac{3n-1}{n+1} = \lim_{n\to\infty} \dfrac{3 - \frac{1}{n}}{1 + \frac{1}{n}} = \dfrac{3-0}{1+0} = 3$ and also

$\lim_{n\to\infty} \dfrac{3n^2+6n+2}{n^2+2n+1} = \lim_{n\to\infty} \dfrac{3 + \frac{6}{n} + \frac{2}{n^2}}{1 + \frac{2}{n} + \frac{1}{n^2}} = 3$ and therefore

$\lim_{n\to\infty} x_n = 3$ by the "sandwich" or "squeeze" theorem.

8-2

Show, using the definition of the limit of a sequence, that $\lim_{n\to\infty} \frac{3}{n} = 0$.

****

We wish to show that for every $\epsilon > 0$ there is a positive integer N such that if $n > N$, $\left|\frac{3}{n} - 0\right| < \epsilon$.

Suppose $\epsilon > 0$. $\frac{3}{\epsilon} > 0$. By the theorem of Archimedes there is a positive integer N such that $N > \frac{3}{\epsilon}$. Now if $n > N$, $n > N > \frac{3}{\epsilon}$, whence, by elementary algebra we have $\frac{1}{n} < \frac{\epsilon}{3}$.

So if $n > N$ then
$$\left|\frac{3}{n} - 0\right| = \frac{3}{n}$$
$$= 3\left(\frac{1}{n}\right)$$
$$= 3\left(\frac{\epsilon}{3}\right)$$
$$= \epsilon .$$

Hence by the definition of limit,
$$\lim_{n\to\infty} \frac{3}{n} = 0.$$

MONOTONIC AND BOUNDED SEQUENCES

8-3

Discuss the boundedness and monotonicity of the sequence whose nth term is $\frac{n}{4n + 1}$.

****

$$\frac{4n^2+5n}{4n^2+5n+1} < 1$$

$$\frac{n(4n+5)}{(n+1)(4n+1)} < 1$$

$$\frac{n}{4n+1} < \frac{n+1}{4n+5}$$

$$\frac{n}{4n+1} < \frac{n+1}{4(n+1)+1}$$

$$\lim_{n\to\infty} \frac{n}{4n+1} = \lim_{n\to\infty} \frac{1}{4+\frac{1}{n}} = \frac{1}{4}$$

THUS, SEQUENCE IS INCREASING
CLEARLY BOUNDED BELOW BY $\frac{1}{5}$
BOUNDED ABOVE BY $\frac{1}{4}$

8-4

The sequence defined by $x_n = (n + 1)(n + 3)/n^2$ is (a) decreasing (b) increasing (c) non-monotonic (d) divergent (e) bounded above but not bounded below.

****

$$X_n = \frac{(n+1)(n+3)}{n^2} = \frac{n+1}{n} \cdot \frac{n+3}{n} = \left(1+\frac{1}{n}\right)\left(1+\frac{3}{n}\right) > \left(1+\frac{1}{n+1}\right)\left(1+\frac{3}{n+1}\right) = X_{n+1},$$

so $\{X_n\}$ is decreasing.

8-5

Consider the recursive sequence defined by

$$x_1 = 1; \quad x_{n+1} = \frac{x_n^2 + 2}{2x_n}, \quad n > 1.$$

1. Evaluate the first three terms of this sequence.
2. You may assume the sequence to be monotone (after the first term) and bounded and hence convergent. Find its limit.

****

1. $x_1 = 1$. $x_2 = \frac{1+2}{2\cdot 1} = \frac{3}{2}$. $x_3 = \frac{\left(\frac{3}{2}\right)^2 + 2}{2\left(\frac{3}{2}\right)} = \frac{\frac{9}{4}+2}{3} = \frac{17}{12}$

2. Let $\lim_{n\to\infty} x_n = L$. Take the limit of both sides of the recursive part of the definition.

$$\lim_{n\to\infty} x_{n+1} = \lim_{n\to\infty} \frac{x_n^2+2}{2x_n}$$

$$L = \frac{L^2+2}{2L}$$

$$2L^2 = L^2 + 2, \quad L^2 = 2, \quad L = \sqrt{2}$$

Take the positive square root in the last line because all terms in the sequence are positive.

8-6

Tell whether the sequence $\{\frac{\sin(n)}{\sqrt{n}}\}$ converges, and, if so, give its limit.

****

It converges to 0. (The numerator is between -1 and 1, and the denominator tends to infinity.)

INFINITE SERIES OF POSITIVE TERMS

8-7

Establish the convergence or divergence of the series $\sum_{n=1}^{\infty} \frac{(n+3)!-n!}{2^n}$

****

FIRST NOTE $(n+3)! = (n+3)(n+2)(n+1)\,n!$

SO $(n+3)! - n! = n!\,[(n+3)(n+2)(n+1) - 1]$

LET $a_n = \frac{(n+3)!-n!}{2^n}$ AND APPLY THE RATIO TEST

$$\frac{a_{n+1}}{a_n} = \frac{(n+1)!\,[(n+4)(n+3)(n+2)-1]}{2^{n+1}} \cdot \frac{2^n}{n!\,[(n+3)(n+2)(n+1)-1]}$$

$$= \frac{n}{2}\left[\frac{(n+4)(n+3)(n+2)-1}{(n+3)(n+2)(n+1)-1}\right] \geqslant \frac{n}{2}$$

SINCE $\frac{n}{2} \to \infty$ as $n \to \infty$ WE HAVE

$\frac{a_{n+1}}{a_n} \to \infty$ AS $n \to \infty$ SO

$\sum a_n$ DIVERGES

8-8

Compare the series $\sum_{n=1}^{\infty} \frac{n + \ln(n)}{n^3 + n + 1}$ and $\sum_{n=1}^{\infty} \frac{n + \ln(n)}{n^2 + 1}$.

****

Clearly $n > \ln(n)$ and therefore the left series, $\sum_{n=1}^{\infty} \frac{n+\ln(n)}{n^3+n+1}$, n^{Th} term is dominated by $\frac{2n}{n^3+n+1}$. Since $\lim_{n\to\infty} \frac{\frac{2n}{n^3+n+1}}{\frac{1}{n^2}} = \lim_{n\to\infty} \frac{2n^3}{n^3+n+1} = 2 > 0$ so we can conclude from the Limit Comparison Test that the series $\sum_{n=1}^{\infty} \frac{2n}{n^3+n+1}$ is convergent. Now that the series $\sum_{n=1}^{\infty} \frac{2n}{n^3+n+1}$ is a positive term convergent series and $\frac{2n}{n^3+n+1} > \frac{n+\ln(n)}{n^3+n+1}$ we can conclude from the Comparison Test that $\sum_{n=1}^{\infty} \frac{n+\ln(n)}{n^3+n+1}$ is convergent.

The convergence or divergence of a series is not changed with the removal of the 1ST two terms.

Therefore the series $\sum_{n=1}^{\infty} \frac{n+\ln(n)}{n^2+1}$ and $\sum_{n=3}^{\infty} \frac{n+\ln(n)}{n^2+1}$ are both convergent or divergent.

For $n \geq 3$ we have $\ln(n) > 1$, thus $\frac{n+\ln(n)}{n^2+1} > \frac{n+1}{n^2+1}$. Since $\lim_{n\to\infty} \frac{\frac{n+1}{n^2+1}}{\frac{1}{n}} = 1 > 0$

we conclude the positive term series $\sum_{n=3}^{\infty} \frac{n+1}{n^2+1}$ is divergent because of the Limit Comparison Test. Hence we can conclude that $\sum_{n=3}^{\infty} \frac{n+\ln(n)}{n^2+1}$ is divergent from the Comparison test and therefore $\sum_{n=1}^{\infty} \frac{n+\ln(n)}{n^2+1}$ is divergent.

8-9

Find the sum of the following series: $\sum_{n=1}^{\infty} 2^n/5^{n+1}$

****

$\sum_{n=1}^{\infty} \frac{2^n}{5^{n+1}} = \frac{2}{5^2} + \frac{4}{5^3} + \frac{8}{5^4} + \cdots$ is geometric with $a = \frac{2}{25}$ and $r = \frac{2}{5}$. Since $|r|<1$, it converges to

$$\frac{a}{1-r} = \frac{\frac{2}{25}}{3/5} = \frac{2}{25} \cdot \frac{5}{3} = \frac{2}{15}.$$

8-10

Determine whether $\sum_{n=1}^{\infty} \frac{\cos(n) + 3^n}{n^2 + 5^n}$ is convergent or divergent.

****

Note that this is a positive term series since $\cos(n) \in [-1, 1]$. I will show that the given series is bounded from above by a convergent geometric series and hence by the comparison test the given series is convergent.

$$\frac{\cos n + 3^n}{n^2 + 5^n} \le \frac{1 + 3^n}{n^2 + 5^n} < \frac{3^n + 3^n}{n^2 + 5^n} < \frac{2(3)^n}{5^n} = \left(\frac{3}{5}\right)^n (2).$$

The product of a finite number and a convergent series remains convergent.

ALTERNATING SERIES AND ABSOLUTE CONVERGENCE

8-11

Which of the following series is convergent but not absolutely convergent?
(a) $\sum_{n=1}^{\infty} 1/n$ (b) $\sum_{n=1}^{\infty} \frac{\sin n}{n^2}$ (c) $\sum_{n=1}^{\infty} \frac{(-1)^n}{\sqrt{n}}$ (d) $\sum_{n=1}^{\infty} \frac{3^n}{2^n + \sqrt{n}}$ (e) $\sum_{n=1}^{\infty} \frac{1 - 2n}{n + 1}$.

****

$\sum_{n=1}^{\infty} \frac{(-1)^n}{\sqrt{n}}$ is convergent by the alternating series test but not absolutely convergent since $\sum_{n=1}^{\infty} \left| \frac{(-1)^n}{\sqrt{n}} \right| = \sum_{n=1}^{\infty} \frac{1}{n^{1/2}}$ is a divergent p-series.

8-12

Determine if the following infinite series converges absolutely, converges conditionally, or diverges:

$$\sum_{k=1}^{\infty} \frac{(-1)^k k}{k^2 + 1}$$

****

Since this is an alternating series, and

$$\lim_{k\to\infty} \frac{k}{k^2+1} = 0,$$ then the series must converge.

$$\sum_{k=1}^{\infty} \left| \frac{(-1)^k k}{k^2+1} \right| = \sum_{k=1}^{\infty} \frac{k}{k^2+1}$$ diverges since it is asymptotically proportional to a divergent series, the harmonic series:

$$\lim_{k\to\infty} \frac{\frac{k}{k^2+1}}{1/k} = \lim_{k\to\infty} \frac{k^2}{k^2+1} = 1$$

∴ The given series converges conditionally.

8-13

Determine whether the following series is conditionally convergent, absolutely convergent or divergent.

$$\sum_{n=1}^{+\infty} \frac{(-1)^{n+1}}{\sqrt{n}}$$

****

Apply the ratio test for absolute convergence.

$$\lim_{n\to+\infty}\left|\frac{u_{n+1}}{u_n}\right| = \lim_{n\to+\infty}\frac{\frac{1}{\sqrt{n+1}}}{\frac{1}{\sqrt{n}}} = \lim_{n\to+\infty}\frac{\sqrt{n}}{\sqrt{n+1}} = 1$$

So the ratio test fails. However $\sum_{n=1}^{+\infty}|u_n| = \sum_{n=1}^{+\infty}\frac{1}{\sqrt{n}} = \sum_{n=1}^{+\infty}\frac{1}{n^{1/2}}$ which diverges since it is a p-series, $p = 1/2 < 1$.

$\therefore \sum_{n=1}^{+\infty}\frac{(-1)^{n+1}}{\sqrt{n}}$ is not absolutely convergent.

But it is alternating, so use the alternating test. To be convergent, $|u_{n+1}| \le |u_n|$ and $\lim_{n\to+\infty}|u_n|$ must be zero.

$|u_{n+1}| = \frac{1}{\sqrt{n+1}}$ & $|u_n| = \frac{1}{\sqrt{n}}$, so if $|u_{n+1}| \le |u_n|$,

$\frac{1}{\sqrt{n+1}} \le \frac{1}{\sqrt{n}}$ or $\sqrt{n} \le \sqrt{n+1}$ or $n \le n+1$

which is true for $n \ge 1$. Also $\lim_{n\to+\infty}|u_n|$

$= \lim_{n\to+\infty}\frac{1}{\sqrt{n}} = 0$. $\therefore \sum_{n=1}^{+\infty} u_n$ converges.

Thus $\sum_{n=1}^{+\infty}\frac{(-1)^{n+1}}{\sqrt{n}}$ is conditionally convergent.

8-14

Classify the following series as absolutely convergent, conditionally convergent or divergent.

$$\sum_{k=2}^{\infty} \frac{(-1)^{k+1}}{\ell n\ k}$$

****

Consider absolute convergence first.

Now $\left|\frac{(-1)^{K+1}}{\ln K}\right| = \frac{1}{\ln K}$, but comparing this series with the divergent harmonic series by the limit comparison test gives:

$$\lim_{K \to \infty} \left[\frac{1/\ln K}{1/K}\right] = \lim_{K \to \infty} \left[\frac{K}{\ln K}\right]$$

(By L'Hôpital's Rule) $$= \lim_{K \to \infty} \left[\frac{1}{1/K}\right] = +\infty$$

Thus the series does not converge absolutely.

However, $\lim_{K \to \infty} \left[\frac{(-1)^{K+1}}{\ln K}\right] = 0$

and $\left|\frac{(-1)^{K+1}}{\ln K}\right| > \left|\frac{(-1)^{K+2}}{\ln (K+1)}\right|$

so the series converges by the alternating series test.

Therefore, $\sum_{K=2}^{\infty} \frac{(-1)^{K+1}}{\ln K}$ converges conditionally.

8-15

Determine whether $\sum_{n=1}^{\infty} \frac{(-1)^{n+1} n}{n^2 + 1}$ converges absolutely, converges conditionally, or diverges.

****

Absolute convergence: $\sum_{n=1}^{\infty} \left| \frac{(-1)^{n+1} n}{n^2+1} \right| = \sum_{n=1}^{\infty} \frac{n}{n^2+1}$. Use a limit comparison test with $\sum_{n=1}^{\infty} \frac{1}{n}$.

$$\lim_{n\to\infty} \frac{n}{n^2+1} \Big/ \frac{1}{n} = \lim_{n\to\infty} \frac{n^2}{n^2+1} = 1$$

Since $\sum_{n=1}^{\infty} \frac{1}{n}$ diverges, so does $\sum_{n=1}^{\infty} \frac{n}{n^2+1}$, and the original series does not converge absolutely. Conditional convergence: try the alternating series test. Let $a_n = \frac{n}{n^2+1}$ so the series is of the form $\sum_{n=1}^{\infty} (-1)^{n+1} a_n$.

$$\lim_{n\to\infty} a_n = \lim_{n\to\infty} \frac{n}{n^2+1} = 0, \text{ and } a_n > 0 \text{ for all } n.$$

We must show $a_n \geq a_{n+1}$ for all n. Let

$$f(x) = \frac{x}{x^2+1}. \text{ Then } f'(x) = \frac{(x^2+1)(1) - x(2x)}{(x^2+1)^2} = \frac{1-x^2}{(x^2+1)^2} < 0$$

for $x > 1$, so f is a decreasing function for $x > 1$. Then $a_n = f(n) \geq f(n+1) = a_{n+1}$. The series $\sum_{n=1}^{\infty} \frac{(-1)^{n+1} n}{n^2+1}$ converges by the alternating series test, and therefore it converges conditionally, since absolute convergence failed.

8-16

State a typical form of an alternating series and its test for convergence. Approximate the sum of the alternating series given below to within 0.0001.

$$\sum_{k=0}^{\infty} \frac{(-1)^k}{(2k)!} = 1 - \frac{1}{2!} + \frac{1}{4!} - \frac{1}{6!} + \frac{1}{8!} \cdots\cdots\cdots$$

****

An alternating series has terms which are alternately positive and negative. Two typical examples are:

(i) $\sum_{k=1}^{\infty} (-1)^{k+1} a_k = a_1 - a_2 + a_3 - a_4 + \cdots\cdots$ (1)

In this case, the first term is positive.

(ii) $\sum_{k=1}^{\infty} (-1)^{k} a_k = -a_1 + a_2 - a_3 + a_4 - \cdots\cdots$ (2)

In this case, the first term is negative.

The test for convergence stipulates that:

(a) a_k tend to 0; i.e. $\lim_{n \to \infty} a_n = 0$, and

(b) a_k are nonincreasing; i.e.

$$a_1 \geq a_2 \geq a_3 \geq \cdots\cdots a_n \geq a_{n+1} \geq \cdots$$

For the given series, we note that:

the first term $= 1$

the second term $-\frac{1}{2!} = -0.500000$

(1st + 2nd term $= 0.5$)

the third term $\frac{1}{4!} = 0.041666$

(1st + 2nd + 3rd term $= 0.541666$)

the fourth term $-\frac{1}{6!} = -0.001388$

(1st + 2nd + 3rd + 4th term $= 0.540278$)

the fifth term $\frac{1}{8!} = 0.000025$ and so on.

(1st + 2nd + 3rd + 4th + 5th term $= 0.540303$)

Now, the fifth term is the first term less than 0.0001, and it also represents an upper estimate of the error by using the sum of the first four terms as an approximation to the sum.

Hence, we write:

$$\sum_{k=0}^{\infty} \frac{(-1)^k}{(2k)!} \approx 1 - \frac{1}{2!} + \frac{1}{4!} - \frac{1}{6!}$$
$$= 1 - \frac{1}{2} + \frac{1}{24} - \frac{1}{720}$$
$$\approx 0.540278$$

which is accurate to within 0.0001.

POWER SERIES

8-17

Find the power series representation for $\frac{2}{1 - 3x + 3x^2 - x^3}$ where $|x| \leq 1$.

****

Now $\frac{2}{1-3x+3x^2-x^3} = \frac{2}{(1-x)^3} = 2(1-x)^{-3}$. This implies that if $y = (1-x)^{-1} \Rightarrow y' = -1(1-x)^{-2}(-1)$ $\Rightarrow y' = (1-x)^{-2} \Rightarrow y'' = -2(1-x)^{-3}(-1) \Rightarrow y'' = 2(1-x)^{-3}$. Clearly $\frac{1}{1-x}$ can be represented as the geometric series $1 + x + x^2 + \cdots$ where $|x| < 1$. Thus if $z = \frac{1}{1-x} = 1 + x + x^2 + \cdots = \sum_{n=0}^{\infty} x^n, |x| < 1$.

$\Rightarrow z' = \frac{1}{(1-x)^2} = \sum_{n=1}^{\infty} n(x)^{n-1}, |x| < 1 \Rightarrow z'' = \frac{2}{(1-x)^3} = \sum_{n=2}^{\infty} n(n-1)(x)^{n-2}, |x| < 1$. Therefore we have

$$\frac{2}{1-3x+3x^2-x^3} = \frac{2}{(1-x)^3} = \sum_{n=2}^{\infty} n(n-1)(x)^{n-2}, |x| < 1.$$

8-18

Find the interval of convergence of the power series

$$\sum_{k=0}^{\infty} \frac{(-1)^k k^2}{5^k} (x-2)^k$$

****

$$\frac{\left|\frac{(-1)^{k+1}(k+1)^2}{5^{k+1}} (x-2)^{k+1}\right|}{\left|\frac{(-1)^k k^2}{5^k} (x-2)^k\right|} = \left(\frac{k+1}{k}\right)^2 \frac{|x-2|}{5}$$

and the limit is $\frac{|x-2|}{5}$ (as $k \to \infty$).

$\frac{|x-2|}{5} < 1$ when $|x-2| < 5$,

∴ absolutely convergent on $(-3, 7)$ by ratio test

at $x = -3$: $\sum \frac{(-1)^k k^2}{5^k} (-5)^k = \sum k^2$ div.

at $x = 7$: $\sum \frac{(-1)^k k^2}{5^k} (5)^k = \sum (-1)^k k^2$ div.

∴ interval of convergence is $(-3, 7)$

8-19

Find the interval of convergence for:

a. $\sum_{n=1}^{\infty} \frac{x^n}{4n^2}$ b. $\sum_{n=1}^{\infty} \frac{x^n}{n2^n}$

****

a. We apply the ratio test:

$$\lim_{n\to\infty}\left|\frac{u_{n+1}}{u_n}\right| = \lim_{n\to\infty}\left|\frac{\frac{x^{n+1}}{4(n+1)^2}}{\frac{x^n}{4n^2}}\right| = \lim_{n\to\infty}\left|\frac{n^2 x}{(n+1)^2}\right|$$

$$= \lim_{n\to\infty}\frac{n^2}{(n+1)^2}\cdot\lim_{n\to\infty}|x| = 1\cdot|x| = |x|$$

thus we have convergence when $|x|<1$

when $x=1$, we get $\sum_{n=1}^{\infty}\frac{1}{4n^2}$ which is convergent because it is a hyperharmonic series (or p series) with $p=2$.

when $x=-1$, we get $\sum_{n=1}^{\infty}\frac{(-1)^n}{4n^2}$ which is convergent because it is absolutely convergent (as shown when $x=1$)

Therefore the interval of convergence is $[-1, 1]$.

b. $$\lim_{n\to\infty}\left|\frac{u_{n+1}}{u_n}\right| = \lim_{n\to\infty}\left|\frac{\frac{x^{n+1}}{(n+1)2^{n+1}}}{\frac{x^n}{n2^n}}\right| = \lim_{n\to\infty}\frac{n}{n+1}\cdot\left|\frac{x}{2}\right| = \left|\frac{x}{2}\right|$$

thus we have convergence when $\left|\frac{x}{2}\right|<1$ or $|x|<2$.

When $x=2$, we get $\sum_{n=1}^{\infty}\frac{1}{n}$ which is the divergent harmonic.

When $x=-2$, we get $\sum_{n=1}^{\infty}\frac{(-1)^n}{n}$, which is convergent because it is an alternating series where $\lim_{n\to 0} u_n = 0$.

Therefore the interval of convergence is $[-2, 2)$.

8-20

Find the interval of convergence of the following power series.

$$\sum_{n=1}^{+\infty} \frac{(-1)^n (x-2)^n}{\sqrt[3]{n}}$$

****

Applying the ratio test, $\lim_{n \to +\infty} \left| \frac{u_{n+1}}{u_n} \right| =$

$$\lim_{n \to +\infty} \left| \frac{\frac{(x-2)^{n+1}}{\sqrt[3]{n+1}}}{\frac{(x-2)^n}{\sqrt[3]{n}}} \right| = \lim_{n \to +\infty} \left| (x-2) \sqrt[3]{\frac{n}{n+1}} \right|$$

$$= |x-2| \lim_{n \to +\infty} \sqrt[3]{\frac{n}{n+1}} = |x-2| \cdot 1 = |x-2|$$

To converge absolutely $|x-2| < 1$ or

$$-1 < x-2 < 1$$

$$1 < x < 3$$

when $x = 3$, $\sum_{n=1}^{+\infty} \frac{(-1)^n (x-2)^n}{\sqrt[3]{n}} = \sum_{n=1}^{+\infty} \frac{(-1)^n (1)^n}{\sqrt[3]{n}} = \sum_{n=1}^{+\infty} \frac{(-1)^n}{\sqrt[3]{n}}$

which is alternating. Check to see if $|u_{n+1}| \le |u_n|$.

$\frac{1}{\sqrt[3]{n+1}} \le \frac{1}{\sqrt[3]{n}}$ or $\sqrt[3]{n} \le \sqrt[3]{n+1}$ or $n \le n+1$

which is true for $n \ge 1$. Also $\lim_{n \to +\infty} |u_n| = \lim_{n \to +\infty} \frac{1}{\sqrt[3]{n}}$

is zero. So the series converges.

when $x = 1$, $\sum_{n=1}^{+\infty} \frac{(-1)^n (x-2)^n}{\sqrt[3]{n}} = \sum_{n=1}^{+\infty} \frac{(-1)^n (-1)^n}{\sqrt[3]{n}} = \sum_{n=1}^{+\infty} \frac{(-1)^{2n}}{\sqrt[3]{n}} =$

$\sum_{n=1}^{+\infty} \frac{1}{\sqrt[3]{n}}$ which diverges because it is a p-series, $p = 1/3 < 1$. Thus the interval of convergence is $(1, 3]$.

8-21

Consider the power series $\sum_{k=1}^{\infty} \frac{(-1)^k}{k} x^k$

(a) Find the radius of convergence.

(b) Determine what happens at the end points (absolute or conditional convergence, or divergence).

****

a) Let r = radius of convergence

$$\lim_{k\to\infty} \left|\frac{a_{k+1}}{a_k}\right| = \left|\frac{(-1)^{k+1}}{k+1} x^{k+1} \cdot \frac{k}{(-1)^k \cdot x^k}\right|$$

$$= \lim_{k\to\infty} \frac{k}{k+1} |x| = |x|$$

By the ratio test, the series converges absolutely for $|x|<1$ and diverges for $|x|>1$

Hence, $r=1$

b) for $x=1$, we have $\sum_{k=1}^{\infty} \frac{(-1)^k}{k}$ which converges conditionally

for $x=-1$, we have $\sum_{k=1}^{\infty} \frac{(-1)^k}{k} \cdot (-1)^k = \sum_{k=1}^{\infty} \frac{1}{k}$,

which diverges

8-22

Find a power series representation for $\frac{\sin(x)\cos(x)}{x}$

****

Since $\sin 2x = 2\sin x\cos x$ we will rewrite $\frac{\sin x\cos x}{x} = \frac{2\sin x\cos x}{2x} = \frac{\sin 2x}{2x}$.

The Maclaurin series for $\sin x$ is $\sum_{n=0}^{\infty}\frac{(-1)^n x^{2n+1}}{(2n+1)!}$ and we may conclude that the series for $\frac{\sin 2x}{2x}$ is $\sum_{n=0}^{\infty}\frac{(-1)^n(2x)^{2n+1}}{(2n+1)!\,2x} =$

$$\sum_{n=0}^{\infty}\frac{(-1)^n(2x)^{2n}}{(2n+1)!}.$$

8-23

Which of the following is impossible for the convergence set of the power series $\sum_{n=0}^{\infty} a_n(x-3)^n$? (a) $0 < x < 6$ (b) $1 \le x < 5$ (c) $2 \le x \le 5$ (d) $2 \le x \le 4$ (e) $-1 < x \le 7$.

****

Since the series uses powers of $x-3$, the interval of convergence must be centered at 3, hence $2 \le x \le 5$ is impossible.

8-24

Find the convergence set of $\displaystyle\sum_{n=1}^{\infty} \frac{(-1)^n(x-5)^n}{n \cdot 3^n}$

****

$$\lim_{n \to \infty} \frac{|a_{n+1}|}{|a_n|} = \lim_{n \to \infty} \frac{|x-5|^{n+1}}{(n+1) \cdot 3^{n+1}} \cdot \frac{n \cdot 3^n}{|x-5|^n} = \lim_{n \to \infty} \frac{n \, |x-5|}{(n+1) \cdot 3}$$

$= \dfrac{|x-5|}{3}$. So the series is absolutely convergent for

$\dfrac{|x-5|}{3} < 1, \quad |x-5| < 3, \quad -3 < x-5 < 3, \quad 2 < x < 8.$

For $x = 2$, $\displaystyle\sum_{n=1}^{\infty} \frac{(-1)^n(x-5)^n}{n \cdot 3^n} = \sum_{n=1}^{\infty} \frac{(-1)^n(-3)^n}{n \cdot 3^n} = \sum_{n=1}^{\infty} \frac{1}{n}$ diverges.

For $x = 8$, $\displaystyle\sum_{n=1}^{\infty} \frac{(-1)^n(x-5)^n}{n \cdot 3^n} = \sum_{n=1}^{\infty} \frac{(-1)^n \cdot 3^n}{n \cdot 3^n} = \sum_{n=1}^{\infty} \frac{(-1)^n}{n}$ a convergent

alternating series. Therefore the convergence set is $2 < x \leq 8$.

8-25

Find the radius of convergence of the series $\displaystyle\sum_{n=1}^{\infty} \frac{3^n(x-2)^{2n+1}}{n!}$

****

To FIND RADIUS of CONVERENGE of $\sum a_n x^n$
WANT TO USE RATIO TEST.

$$\left| \frac{3^{n+1}(x-2)^{2n+3}}{(n+1)!} \cdot \frac{n!}{3^n(x-2)^{2n+1}} \right| =$$

$$= \frac{3}{(n+1)} |x-2|^2 \longrightarrow 0$$

SO RADIUS of CONVERGENCE $= +\infty$

i.e., THE SERIES CONVERGES FOR ALL REAL x.

DIFFERENTIATION AND INTEGRATION OF POWER SERIES

8-26

Use a Maclaurin series to approximate $\int_0^1 e^{-t^2}\,dt$ with an accuracy of 0.01.

****

$$e^x = 1 + x + \frac{x^2}{2} + \frac{x^3}{6} + \frac{x^4}{24} + \ldots$$

Set $x = -t^2$ to get

$$e^{-t^2} = 1 - t^2 + \frac{t^4}{2} - \frac{t^6}{6} + \frac{t^8}{24} - \ldots$$

$$\int_0^1 e^{-t^2}dt = \int_0^1 \left(1 - t^2 + \frac{t^4}{2} - \frac{t^6}{6} + \frac{t^8}{24} - \ldots\right) dt$$

$$= \left(t - \frac{t^3}{3} + \frac{t^5}{10} - \frac{t^7}{42} + \frac{t^9}{216} - \ldots\right)\Bigg|_0^1$$

$$= 1 - \frac{1}{3} + \frac{1}{10} - \frac{1}{42} + \frac{1}{216} - \ldots$$

(alt. series – need only 4 terms for 0.01 accuracy)

$$\approx 1 - \frac{1}{3} + \frac{1}{10} - \frac{1}{42} \approx 0.7428571$$

8-27

In terms of powers of x, the power series for $\frac{1}{1-x} = \sum_{n=0}^{\infty} x^n$. Find the power series for $\frac{1}{(1-x)^2}$ in terms of powers of x.

****

BY DIFFERENTIATING THE GIVEN SERIES, WE OBTAIN:

$$\frac{1}{(1-x)^2} = \sum_{n=0}^{\infty} n x^{n-1}$$

BUT THIS IS 0 WHEN N = 0.

$$\therefore \frac{1}{(1-x)^2} = \sum_{n=1}^{\infty} n x^{n-1}$$

8-28

Let $f(x) = \sum_{n=0}^{\infty} a_n x^n$. If $g'(x) = f(x)$ and $g(0) = 3$, then $g(x)$ is

(a) $3 + \sum_{n=1}^{\infty} \frac{a_{n-1}}{n} x^n$ (b) $\sum_{n=0}^{\infty} a_n/(n+1)\, x^{n+1}$ (c) $\sum_{n=1}^{\infty} n \cdot a_n \cdot x^{n-1}$ (d) $3 + \sum_{n=1}^{\infty} a_n x^n$

(e) $3 - a_1 + \sum_{n=1}^{\infty} n \cdot a_n \cdot x^{n-1}$.

****

$$g'(x) = f(x) \Rightarrow g(x) = \int f(x)\,dx = \sum_{n=0}^{\infty} \frac{a_n x^{n+1}}{n+1} + C$$

$$= \sum_{n=1}^{\infty} \frac{a_{n-1} x^n}{n} + C. \quad g(0) = 3 \Rightarrow \sum_{n=1}^{\infty} \frac{a_{n-1} \cdot 0^n}{n} + C = 3 \Rightarrow$$

$$C = 3, \text{ so } g(x) = 3 + \sum_{n=1}^{\infty} \frac{a_{n-1} x^n}{n}.$$

TAYLOR SERIES

8-29

Find the Taylor series for xcosx about the origin.

****

$$\cos x = 1 - \frac{x^2}{2} + \frac{x^4}{4!} - \frac{x^6}{6!} + - \cdots$$

$$= \sum_{n=0}^{\infty} \frac{(-1)^n x^{2n}}{(2n)!}, \text{ so}$$

$$x\cos x = x - \frac{x^3}{2} + \frac{x^5}{4!} - \frac{x^7}{7!} + - \cdots$$

$$= \sum_{n=0}^{\infty} \frac{(-1)^n x^{2n+1}}{(2n)!}$$

8-30

Give the Taylor Series expansion of f(x) = sin(x) at the point c = π/4. (Note: sin(π/4) = cos(π/4) = (√2)/2).

****

$$\sum_{i=0}^{\infty} \frac{f^{(i)}(c)}{i!}(x-c)^i = \sin\frac{\pi}{4} + \cos\frac{\pi}{4}\cdot\left(x-\frac{\pi}{4}\right) - \sin\frac{\pi}{4}\cdot\frac{1}{2!}\left(x-\frac{\pi}{4}\right)^2$$

$$- \cos\frac{\pi}{4}\cdot\frac{1}{3!}\left(x-\frac{\pi}{4}\right)^3 + \sin\frac{\pi}{4}\cdot\frac{1}{4!}\left(x-\frac{\pi}{4}\right)^4 + \cdots$$

$$= \frac{\sqrt{2}}{2} + \frac{\sqrt{2}}{2}\cdot\left(x-\frac{\pi}{4}\right) - \frac{\sqrt{2}}{2}\cdot\frac{1}{2!}\left(x-\frac{\pi}{4}\right)^2 - \frac{\sqrt{2}}{2}\cdot\frac{1}{3!}\left(x-\frac{\pi}{4}\right)^3 + \cdots.$$

8-31

Obtain the Taylor's series development of the given function about the point a = 2.

$$y = \ln x$$

****

$f(x) = \ln x \qquad f(2) = \ln 2$

$f'(x) = x^{-1} \qquad f'(2) = \frac{1}{2}$

$f''(x) = -x^{-2} \qquad f''(2) = -\frac{1}{2^2}$

$f'''(x) = 1\cdot 2x^{-3} \qquad f'''(2) = \frac{1\cdot 2}{2^3}$

$f^4(x) = -1\cdot 2\cdot 3\, x^{-4} \qquad f^4(2) = -\frac{1\cdot 2\cdot 3}{2^4}$

$\vdots \qquad \vdots$

$f^n(x) = (-1)^{n-1}(n-1)!\, x^{-n}, \qquad f^n(2) = \frac{(-1)^{n-1}(n-1)!}{2^n}$

Therefore,

$$\ln x = \ln 2 + \frac{1}{2}(x-2) - \frac{1}{2!\,2^2}(x-2)^2 + \frac{2!}{3!\,2^3}(x-2)^3 - \cdots + \frac{(-1)^{n-1}(n-1)!}{n!\,2^n}(x-2)^n + \cdots$$

$$= \ln 2 + \left(\frac{x-2}{2}\right) - \frac{1}{2}\left(\frac{x-2}{2}\right)^2 + \frac{1}{3}\left(\frac{x-2}{2}\right)^3 - \cdots$$

$$= \ln 2 + \sum_{n=1}^{\infty} \frac{(-1)^{n-1}}{n}\left(\frac{x-2}{2}\right)^n$$

8-32

Find the Taylor polynomial of degree 4 at 0 for the function defined by

$$f(x) = \ln(1 + x).$$

Then compute the value of ln(1.1) accurate to as many decimal places as the polynomial of degree 4 allows.

****

$f(x) = \ln(1+x)$, $a = 0$ and $n = 4$

$f(x) = \ln(1+x) \qquad f(a) = f(0) = \ln(1+0) = \ln(1) = 0$

$f'(x) = \frac{1}{1+x} = (1+x)^{-1} \qquad f'(a) = f'(0) = (1+0)^{-1} = 1$

$f''(x) = -(1+x)^{-2} \qquad f''(a) = f''(0) = -(1+0)^{-2} = -1$

$f'''(x) = 2(1+x)^{-3} \qquad f'''(a) = f'''(0) = 2(1+0)^{-3} = 2$

$f^{IV}(x) = -6(1+x)^{-4} \qquad f^{IV}(a) = f^{IV}(0) = -6(1+0)^{-4} = -6$

$$\therefore \ln(1+x) \approx f(a) + \frac{f'(a)(x-a)}{1!} + \frac{f''(a)(x-a)^2}{2!} + \frac{f'''(a)(x-a)^3}{3!} + \frac{f^{IV}(a)(x-a)^4}{4!}$$

$$\ln(1+x) \approx 0 + \frac{1}{1}x + \frac{(-1)}{2}x^2 + \frac{2}{6}x^3 + \frac{(-6)}{24}x^4$$

or

$$\ln(1+x) = x - \tfrac{1}{2}x^2 + \tfrac{1}{3}x^3 - \tfrac{1}{4}x^4$$

Since $\ln(1.1) = \ln(1+.1)$, let $x = 0.1$ and substitute into the above polynomial.

$$\ln(1.1) \approx (.1) - \tfrac{1}{2}(.1)^2 + \tfrac{1}{3}(.1)^3 - \tfrac{1}{4}(.1)^4$$

$$= 0.1 - .005 + .000333\cdots - .000025$$

$\ln(1.1) \approx .095$, accurate to three decimal places.

8-33

a) Derive the Taylor series about $x = 0$ for $f(x) = e^x$.

b) Use the result of part (a) to obtain the series expansion for e^{-x^2}.

c) Use the result of part (b) to obtain $\int_0^1 e^{-x^2}\,dx$ to two decimal place accuracy.

****

a) For $f(x) = e^x$ we have

$$f'(x) = f''(x) = \cdots = f^{(n)}(x) = e^x$$

and $f'(0) = f''(0) = \cdots = f^{(n)}(0) = 1$

Thus,

$$f(x) = e^x = \sum_{K=0}^{\infty} \frac{f^{(K)}(0)}{K!} x^K = \underline{\underline{\sum_{K=0}^{\infty} \frac{x^K}{K!}}}$$

b) Since $e^{-x^2} = f(-x^2)$, we have

$$e^{-x^2} = \sum_{K=0}^{\infty} \frac{(-x^2)^K}{K!} = \underline{\underline{1 - x^2 + \frac{x^4}{2!} - \frac{x^6}{3!} + \frac{x^8}{4!} - \cdots}}$$

c) Now

$$\int_0^1 e^{-x^2}\,dx = \int_0^1 \left(1 - x^2 + \frac{x^4}{2!} - \frac{x^6}{3!} + \cdots\right) dx$$

$$= \left(x - \frac{x^3}{3} + \frac{x^5}{5(2!)} - \frac{x^7}{7(3!)} + \cdots\right)\Big]_0^1$$

$$= 1 - 1/3 + 1/10 - 1/42 + 1/216 - \cdots$$

Note that this is an alternating series and that $1/216 < .005 = .5 \times 10^{-2}$.

So

$$\int_0^1 e^{-x^2}\,dx \approx 1 - \frac{1}{3} + \frac{1}{10} - \frac{1}{42} = \underline{\underline{.74}}$$

THE BINOMIAL SERIES

8-34

Express $\dfrac{1}{\sqrt{1 + x}}$ as a power series in x and from the result obtain a binomial series for $\dfrac{1}{\sqrt{1 - x^2}}$. Show the result of its integration.

****

When $|x| < 1$, we have from the binomial theorem the following power series for the expression $(1 + x)^m$ where m is not a positive integer:

$$1 + mx + \frac{m(m-1)}{2!}(x^2) + \frac{m(m-1)(m-2)}{3!}(x^3) + \cdots\cdots + \frac{m(m-1)(m-2)\cdots\cdots(m-n+1)}{n!}(x^n) \quad \cdots\cdots (1)$$

For $\dfrac{1}{\sqrt{1+x}} = (1+x)^{-1/2}$ we get from Eq. (1)

$$(1+x)^{-1/2} = 1 + \left(-\frac{1}{2}\cdot x\right) + \frac{\left(-\frac{1}{2}\right)\left(-\frac{1}{2}-1\right)}{2!}(x^2) + \frac{\left(-\frac{1}{2}\right)\left(-\frac{1}{2}-1\right)\left(-\frac{1}{2}-2\right)}{3!}(x^3) + \cdots\cdots + \frac{\left(-\frac{1}{2}\right)\left(-\frac{1}{2}-1\right)\left(-\frac{1}{2}-2\right)\cdots\cdots\left(-\frac{1}{2}-n+1\right)}{n!}(x^n) + \cdots\cdots$$

$$= 1 - \frac{1}{2}x + \frac{1\cdot 3}{2^2\cdot 2!}x^2 - \frac{1\cdot 3\cdot 5}{2^3\cdot 3!}x^3 + (-1)^n\frac{1\cdot 3\cdot 5\cdots\cdots(2n-1)}{2^n\cdot n!}x^n + \cdots\cdots \quad \cdots(2)$$

For $\dfrac{1}{\sqrt{1-x^2}} = (1-x^2)^{-1/2}$ we replace x by $(-x^2)$ in Eq. (2) and write:

$$(1-x^2)^{-1/2} = 1 + \frac{1}{2}x^2 + \frac{1\cdot 3}{2^2\cdot 2!}x^4 + \frac{1\cdot 3\cdot 5}{2^3\cdot 3!}x^6 + \cdots\cdots + \frac{1\cdot 3\cdot 5\cdots\cdots(2n-1)}{2^n\cdot n!}x^{2n} + \cdots\cdots \quad \cdots\cdots (3)$$

By applying the theorem for term by term integration of a power series to Eq. (3) we get:

$$\int_0^x \frac{dt}{\sqrt{1-t^2}} = x + \frac{1}{2}\cdot\frac{x^3}{3} + \frac{1\cdot 3}{2^2\cdot 2!}\cdot\frac{x^5}{5}$$

$$+ \frac{1\cdot 3\cdot 5}{2^3\cdot 3!}\cdot\frac{x^7}{7} + \dots + \frac{1\cdot 3\cdot 5\cdot\ldots\cdot(2n-1)}{2^n\cdot n!}\cdot\frac{x^{2n+1}}{2n+1}$$

$$+ \dots\dots$$

Ans.

8-35

Use the binomial series formula to obtain the Maclaurin series for $f(x) = (1 + x)^{1/3}$.

****

$$(1+x)^{\frac{1}{3}} = 1 + \tfrac{1}{3}x + \frac{\frac{1}{3}\left(-\frac{2}{3}\right)x^2}{2!} + \frac{\frac{1}{3}\left(-\frac{2}{3}\right)\left(-\frac{5}{3}\right)x^3}{3!} + \frac{\frac{1}{3}\left(-\frac{2}{3}\right)\left(-\frac{5}{3}\right)\left(-\frac{8}{3}\right)x^4}{4!} + \dots$$

$$= 1 + \tfrac{1}{3}x + \sum_{n=2}^{\infty} (-1)^{n+1}\frac{2\cdot 5\cdot 8\cdots(3n-4)}{3^n\cdot n!}x^n, \quad \text{for } |x|<1.$$

8-36

Find a power series representation for $(b + x)^{3/2}$ where b is a perfect square, and state the radius of convergence in terms of b.

****

We will use the Binomial Theorem to find the wanted series.

$$(b+x)^{3/2} = b^{3/2} + \frac{\left(\frac{3}{2}\right) b^{1/2} x}{1} + \frac{\left(\frac{3}{2}\right)\left(\frac{1}{2}\right) b^{-\frac{1}{2}} x^2}{2!} + \frac{\left(\frac{3}{2}\right)\left(\frac{1}{2}\right)\left(-\frac{1}{2}\right) b^{-3/2} x^3}{3!}$$

$$+ \frac{\left(\frac{3}{2}\right)\left(\frac{1}{2}\right)\left(-\frac{1}{2}\right)\left(-\frac{3}{2}\right) b^{-5/2} x^4}{4!} + \cdots$$

$$= b^{3/2} + \frac{3b^{1/2}x}{2} + \frac{3b^{-1/2}x^2}{2^2 \cdot 2!} - \frac{3b^{-3/2}x^3}{2^3 \cdot 3!} +$$

$$\frac{3 \cdot 3\, b^{-5/2} x^4}{2^4 \cdot 4!} - \frac{3 \cdot 3 \cdot 5\, b^{-7/2} x^5}{2^5 \cdot 5!} + \frac{3 \cdot 3 \cdot 5 \cdot 7\, b^{-9/2} x^6}{2^6 \cdot 6!} - \cdots$$

$$= b^{3/2} + \frac{3b^{1/2}x}{2} + 3\sum_{n=2}^{\infty} \frac{(-1)^n 1 \cdot 3 \cdot 5 \cdots (2n-5) x^n b^{\frac{3-2n}{2}}}{2^n \cdot n!}.$$

Now to find the radius of convergence.

$$\lim_{n\to\infty} \left| \frac{1 \cdot 3 \cdot 5 \cdots (2n-5)(2n-3)\, x^{n+1}\, b^{-\frac{(2n-1)}{2}}}{2^n \cdot 2 \cdot n!\,(n+1)} \cdot \frac{2^n \cdot n!}{1 \cdot 3 \cdot 5 \cdots (2n-5)\, x^n\, b^{\frac{-2n+3}{2}}} \right| < 1$$

$$\Rightarrow \lim_{n\to\infty} \left| \frac{(2n-3)\, x\, b^{-\frac{(2n-1)}{2} + \frac{(2n-3)}{2}}}{2(n+1)} \right| < 1 \Rightarrow b^{-1}|x| < 1$$

$\Rightarrow |x| < b$ ∴ The radius of convergence is b.

THE MACLAURIN SERIES

8-37

Find a Taylor series of degree 4 about x = 0 for f(x) = log sec x.

****

The Taylor series polynomial about $x=a$

$$P_n(x) = f(a) + (x-a)f'(a) + \frac{(x-a)^2}{2!}f''(a) + \frac{(x-a)^3}{3!}f'''(a) + \cdots + \frac{(x-a)^n}{n!}f^n(a)$$

The 4th degree polynomial about $x=0$, yields the Maclaurin series

$$P_4(x) = f(0) + xf'(0) + \frac{x^2 f''(0)}{2!} + \frac{x^3 f'''(0)}{3!} + \frac{x^4 f^{IV}(0)}{4!}$$

we need to find the values of the coefficients $f(0), f'(0) \ldots f^{IV}(0)$.

$$f(x) = \log\sec x \rightarrow f(0) = \log\sec 0 = 0$$

$$f'(x) = \tan x \rightarrow f'(0) = \tan 0 = 0$$

$$f''(x) = \sec^2 x \rightarrow f''(0) = \sec^2 0 = 1$$

$$f'''(x) = 2\sec^2 x\tan x \rightarrow f'''(0) = 2\sec^2 0 . \tan 0 = 0$$

$$f^{IV}(0) = 2\sec^4 x + 4\tan^2 x\sec^2 x \rightarrow f^{IV}(0) = 2\sec^2 0 + 4\tan^2 0\sec^2 0 = 2$$

Substituting these values into $P_4(x)$

$$P_4(x) = 0 + 0x + \frac{1x^2}{2!} + \frac{0x^3}{3!} + \frac{2x^4}{4!}$$

Hence $\log\sec x \sim P_4(x) = \frac{x^2}{2} + \frac{x^4}{12}$

8-38

a) Find the Maclaurin series expansion with n = 5 for $f(x) = 2^x$.
b) Use this expansion to approximate $2^{.1}$.

****

a) Derivatives of exponential functions are most easily carried out when the base is e. Thus we begin by writing

$$f(x) = 2^x = e^{\ln 2^x} = e^{x \ln 2}$$

then

$$f'(x) = (\ln 2)\, e^{x \ln 2}$$
$$f''(x) = (\ln 2)^2\, e^{x \ln 2}$$
$$f'''(x) = (\ln 2)^3\, e^{x \ln 2}$$
$$f^{IV}(x) = (\ln 2)^4\, e^{x \ln 2}$$
$$f^{V}(x) = (\ln 2)^5\, e^{x \ln 2}$$

thus

$$f(x) \cong f(0) + f'(0)x + \frac{f''(0)x^2}{2!} + \cdots + \frac{f^{V}(0)x^2}{5!} + \cdots$$

$$= 1 + (\ln 2)x + \frac{(\ln 2)^2 x^2}{2!} + \frac{(\ln 2)^3 x^3}{3!} + \frac{(\ln 2)^4 x^4}{4!} + \frac{(\ln 2)^5 x^5}{5!} + \cdots$$

b)

$$(2^{.1}) \doteq 1 + (\ln 2)(.1) + \tfrac{1}{2}(\ln 2)^2(.1)^2 + \tfrac{1}{6}(\ln 2)^3(.1)^3 + \tfrac{1}{24}(\ln 2)^4(.1)^4$$

$$= \boxed{1.0718}$$

(by calculator)

8-39

Find the Maclaurin series expansion for f(x)= ln(1-x) and determine the interval of convergence.

****

The Maclaurin series expansion for $f(x)$ is given by $f(x) = f(0) + \frac{f'(0)}{1!}x + \frac{f''(0)}{2!}x^2$ $\frac{f'''(0)}{3!}x^3 + \cdots + \frac{f^n(0)}{n!}x^n + \cdots$

Now $f(x) = \ln(1-x)$ so $f(0) = 0$

$f'(x) = -\frac{1}{1-x}$ so $f'(0) = -1$

$f''(x) = -\frac{1}{(1-x)^2}$ so $f''(0) = -1$

$f'''(x) = \frac{-2}{(1-x)^3}$ so $f'''(0) = -2$

$f^{iv}(x) = -\frac{6}{(1-x)^4}$ so $f^{iv}(0) = -6$

$\cdots$

$f^{(n)}(x) = -\frac{(n-1)!}{(1-x)^n}$ so $f^{(n)}(0) = -(n-1)!$

Hence $f(x) = \sum_{n=1}^{\infty} \frac{-(n-1)!}{n!} x^n = \sum_{n=1}^{\infty} -\frac{1}{n} x^n$

By the ratio test $\sum_{n=1}^{\infty} -\frac{1}{n}x^n$ converges

when $\lim_{n\to\infty} \left| \frac{-\frac{1}{n+1}x^{n+1}}{\frac{1}{n}x^n} \right| < 1$; hence, $-1 < x < 1$ the series $\sum_{n=1}^{\infty} -\frac{1}{n}x^n$

Converges. Furthermore $\sum_{n=1}^{\infty} -\frac{1}{n}x^n$

Converges by the alternating series test when $x = -1$, however, it diverges when $x = 1$. So the series $\sum_{n=1}^{\infty} -\frac{1}{n}x^n$ converges to $\ln(1-x)$ for all x such that $-1 \le x < 1$.

8-40

Find and simplify Maclaurin's formula with remainder for $f(x) = e^{3x}$, $n = 4$.

****

$f(x) = e^{3x}$, $f'(x) = 3e^{3x}$, $f^{(2)}(x) = 9e^{3x}$, $f^{(3)}(x) = 27e^{3x}$, $f^{(4)}(x) = 81e^{3x}$,

$f^{(5)}(x) = 243e^{3x}$.

$f(0) = 1$, $f'(0) = 3$, $f^{(2)}(0) = 9$, $f^{(3)}(0) = 27$, $f^{(4)}(0) = 81$.

$$f(x) = f(0) + f'(0)x + \frac{f^{(2)}(0)}{2!}x^2 + \frac{f^{(3)}(0)}{3!}x^3 + \frac{f^{(4)}(0)}{4!}x^4 + \frac{f^{(5)}(z)}{5!}x^5$$

$$e^{3x} = 1 + 3x + \frac{9}{2}x^2 + \frac{27}{6}x^3 + \frac{81}{24}x^4 + \frac{243e^{3z}}{120}x^5$$

$$e^{3x} = 1 + 3x + \frac{9}{2}x^2 + \frac{9}{2}x^3 + \frac{27}{8}x^4 + \frac{81e^{3z}}{40}\cdot x^5,$$ where z is between 0 and x.

8-41

Write the Taylor Polynomial at 0 of degree 4 for $f(x) = \ln(1+x)$.

****

$f(x) = \ln(1+x)$ $\qquad f(0) = \ln 1 = 0$

$f^{(1)}(x) = \frac{1}{1+x} = (1+x)^{-1}$ $\qquad f^{(1)}(0) = 1$

$f^{(2)}(x) = -(1+x)^{-2}$ $\qquad f^{(2)}(0) = -1$

$f^{(3)}(x) = 2(1+x)^{-3}$ $\qquad f^{(3)}(0) = 2$

$f^{(4)}(x) = -6(1+x)^{-4}$ $\qquad f^{(4)}(0) = -6$

THE TAYLOR POLYNOMIAL OF DEGREE 4 AT 0 IS

$$T(x) = \frac{f(0)}{0!} + \frac{f^{(1)}(0)}{1!}x + \frac{f^{(2)}(0)}{2!}x^2 + \frac{f^{(3)}(0)}{3!}x^3 + \frac{f^{(4)}(0)}{4!}x^4$$

$$= \frac{0}{1} + \frac{1}{1}x + \frac{-1}{2}x^2 + \frac{2}{6}x^3 + \frac{-6}{24}x^4$$

$$= x - \frac{1}{2}x^2 + \frac{1}{3}x^3 - \frac{1}{4}x^4.$$

8-42

If the Maclaurin series for $f(x)$ is $1 - 9x + 16x^2 - 25x^3 + \cdots$, then $f^{(3)}(0)$ is equal to (a) -25 (b) $-25/6$ (c) -150 (d) $-25/3$ (e) -75.

****

The coefficient of X^3 is $\frac{f^{(3)}(0)}{3!}$, hence $\frac{f^{(3)}(0)}{6} = -25$

so $f^{(3)}(0) = 6(-25) = -150$.

8-43

Use the first three non-zero terms of a MacLaurin's Series to estimate
$\int_0^1 e^{-x^2}\,dx$

****

$$e^u \approx 1 + u + \frac{u^2}{2}$$

hence: $e^{-x^2} \approx 1 - x^2 + \frac{x^4}{2}$

thus: $\int_0^1 e^{-x^2}\,dx \approx \int_0^1 dx - \int_0^1 x^2\,dx + \int_0^1 \frac{x^4}{2}\,dx$

$$= x\Big]_0^1 - \frac{1}{3}x^3\Big]_0^1 + \frac{1}{10}x^5\Big]_0^1$$

$$= 1 - \frac{1}{3} + \frac{1}{10} = \frac{30 - 10 + 3}{30} = \frac{23}{30}$$

GEOMETRIC SERIES

8-44

Briefly state the general form of a geometric series and its properties for convergence and divergence. Show the geometric series included in a nonterminating decimal like 0.7888888...... and indicate the rational number represented by this decimal.

****

The general form of a geometric series is given by:

$$1 + x + x^2 + x^3 + \cdots = \sum_{k=0}^{\infty} x^k$$

If $|x| < 1$, the series converges, otherwise, it diverges.

The given number, $0.7888888\ldots$ can be written as:

$$0.7 + 0.08 + 0.008 + 0.0008 + \cdots$$

$$= 0.7 + 0.08\,(1 + 0.1 + 0.001 + 0.0001 + \cdots)$$

$$= \frac{7}{10} + \frac{8}{100}\left(1 + \frac{1}{10} + \frac{1}{10^2} + \frac{1}{10^3} + \cdots\right)$$

$$= \frac{7}{10} + \frac{8}{100} \cdot \frac{1}{1 - \frac{1}{10}}$$

$$= \frac{7}{10} + \frac{8}{100} \cdot \frac{10}{9}$$

$$= \frac{7}{10} + \frac{8}{90} = \frac{71}{90}$$

Hence, the geometric series included in the number $0.7888888\ldots\ldots$ is:

$$\frac{8}{100} + \left(1 + \frac{1}{10} + \frac{1}{10^2} + \frac{1}{10^3} + \cdots\right)$$

and the rational number represented by the decimal is $\frac{71}{90}$.

8-45

$2/9 - 4/27 + \cdots + \dfrac{(-1)^{n+1} \cdot 2^n}{3^{n+1}} + \cdots$ is equal to (a) 2/15 (b) 1/5 (c) 1/6 (d) 21/110 (e) 1/7 .

****

It's geometric with $a = 2/9$ and $r = -\frac{2}{3}$. Since $|r|<1$, the series converges with sum $\frac{a}{1-r} = \frac{2/9}{5/3} = \frac{2}{9} \cdot \frac{3}{5} = \frac{2}{15}$.

8-46

Let $S = \sum_{n=1}^{\infty} \frac{1}{2^{n-1}}$. Determine whether this series converges or diverges and if possible find S.

****

The series is a geometric series with $r = \frac{1}{2} < 1$; hence, the series converges and

$$S = \frac{1}{1-r} = \frac{1}{1-\frac{1}{2}} = \frac{1}{\frac{1}{2}} = 2$$

8-47

Find the value of

$$\sum_{n=2}^{\infty} \frac{3^n + 5^n}{15^n}$$

****

$$\sum_{n=2}^{\infty} \frac{3^n + 5^n}{15^n} = \sum_{n=2}^{\infty} \frac{3^n}{15^n} + \sum_{n=2}^{\infty} \frac{5^n}{15^n}$$

$$= \sum_{n=2}^{\infty} \left(\frac{1}{5}\right)^n + \sum_{n=2}^{\infty} \left(\frac{1}{3}\right)^n = S_1 + S_2$$

Both of these infinite sums are infinite geometric series

The sum of an infinite geometric series $= \frac{a}{1-r}$ if $|r| < 1$

$$\Rightarrow S_1 = \frac{1/25}{1-1/5} = \frac{1/25}{4/5} = \frac{1}{20}$$

where $a = \frac{1}{25}$; $r = \frac{1}{5}$.

$$S_2 = \frac{1/9}{1-1/3} = \frac{1/9}{2/3} = \frac{1}{6} \text{ where } a = \frac{1}{9} ; r = \frac{1}{3}$$

$$\Rightarrow S = S_1 + S_2 = \frac{1}{20} + \frac{1}{6} = \frac{13}{60} .$$

THE INTEGRAL TEST

8-48

Use the integral test to determine if the following series converges or diverges:

$$\sum_{n=1}^{\infty} \frac{n}{(n^2+1)^2}$$

****

$f(x) = \dfrac{x}{(x^2+1)^2}$ is the function to be considered on $[1,\infty)$.

$f(x)$ is positive on $[1,\infty)$, since x is positive there and $(x^2+1)^2$ is always positive.

$f(x)$ is decreasing on $[1,\infty)$, since $f'(x) = \dfrac{(x^2+1)^2 - x\,2(x^2+1)\,2x}{(x^2+1)^4}$

$= \dfrac{1-3x^2}{(x^2+1)^3}$ is negative on $[1,\infty)$

$f(x)$ is continuous on $[1,\infty)$, since $f(x)$ is a rational function and is defined at all points of $[1,\infty)$.

$\therefore$ Conditions for integral test hold.

$$\int_1^{\infty} \frac{x}{(x^2+1)^2}\,dx \quad \begin{pmatrix} u = x^2+1 \\ du = 2x\,dx \end{pmatrix} \quad = \int \frac{\frac{1}{2}du}{u^2} = -\frac{1}{2}\,u^{-1}\Big| = -\frac{1}{2}\left[\frac{1}{x^2+1}\right]_1^{\infty}$$

$$= -\frac{1}{2}\left[\lim_{x\to\infty}\frac{1}{x^2+1} - \frac{1}{1^2+1}\right] = -\frac{1}{2}\left[0 - \frac{1}{2}\right] = 1/4,$$ which is finite. Since the integral has a finite value, the series converges.

8-49

Use the integral test to show that the series

$$\sum_{k=2}^{\infty} \frac{1}{k(\ell n\ k)^p}$$

converges if $p > 1$ and diverges if $p \leqslant 1$.

[Hint: Consider two cases $p = 1$ and $p \neq 1$.]

****

Suppose that $p = 1$. Using the integral test with

$$\int_2^{\infty} \frac{dx}{x \ln k} \quad \text{gives} \quad \lim_{\ell \to +\infty} \left[\ln(\ln x)\right]_2^{\ell} = +\infty$$

and the series diverges.

If $p \neq 1$, then using the integral test with

$$\int_2^{\infty} \frac{dx}{x(\ln x)^p} \quad \text{gives} \quad \lim_{\ell \to +\infty} \left[\frac{(\ln x)^{1-p}}{1-p}\right]_2^{\ell}$$

Since $\ln x \to +\infty$ as $x \to +\infty$, the convergence of the series depends on whether $\ln x$ is in the numerator or denominator of the limit above.

If $p > 1$ then $\lim_{\ell \to +\infty} \left[\frac{(\ln x)^{1-p}}{1-p}\right]_2^{\ell} = \frac{-1}{(1-p)(\ln 2)^{p-1}}$

and the series converges.

If $p < 1$ then $\lim_{\ell \to +\infty} \left[\frac{(\ln x)^{1-p}}{1-p}\right]_2^{\ell} = +\infty$

and the series diverges.

So we have convergence if $p > 1$ and divergence if $p \leq 1$.

8-50

Use the Integral test to determine whether $\sum_{n=2}^{n=\infty} \frac{1}{n(\log n)^4}$ converges or diverges.

****

Using the Integral Test

$\sum_{n=2}^{n=\infty} \frac{1}{n(\log n)^4}$ converges if $\int_2^\infty \frac{1}{x(\log x)^4}\,dx$ converges.

$\sum_{n=2}^{n=\infty} \frac{1}{n(\log n)^4}$ diverges if $\int_2^\infty \frac{1}{x(\log x)^4}\,dx$ diverges

now $\int_2^\infty \frac{1}{x(\log x)^4}\,dx = \lim_{b\to\infty} \int_2^b \frac{1}{x(\log x)^4}\,dx.$

using $u = \log x$ $\quad \frac{du}{dx} = \frac{1}{x} \rightarrow dx = x\,du.$

so $\int \frac{1}{x(\log x)^4}\,dx = \int \frac{1}{x}\frac{1}{u^4}\,x\,du = \int u^{-4}\,du = \frac{u^{-3}}{-3} + c = \frac{-1}{3u^3} + c$

hence $\lim_{b\to\infty} \int_2^b \frac{1}{x(\log x)^4}\,dx = \lim_{b\to\infty} \left[\frac{-1}{3(\log x)^3}\right]_2^b$

$$= \lim_{b\to\infty}\left(\frac{-1}{3(\log b)^3} - \frac{-1}{3(\log 2)^3}\right) = 0 + \frac{1}{3(\log 2)^3}$$

Now since $\int_2^\infty \frac{1}{x(\log x)^4}\,dx$ converges to $\frac{1}{3(\log 2)^3}$ then by the Integral Test $\sum \frac{1}{n(\log n)^4}$ also converges

8-51

Determine whether or not the following infinite series converges:

$$\sum_{k=2}^{\infty} \frac{1}{k(\ln k)^2}$$

****

Use the integral test:

$$\int_2^{\infty} \frac{dx}{x(\ln x)^2} = \lim_{t\to\infty} \int_2^{t} \frac{dx}{x(\ln x)^2} \qquad \text{let } u = \ln x, \quad du = \frac{1}{x}dx$$

$$= \lim_{t\to\infty} \int_{\ln 2}^{\ln t} \frac{du}{u^2} = \lim_{t\to\infty} -\frac{1}{u}\Big|_{\ln 2}^{\ln t}$$

$$= \lim_{t\to\infty}\left[-\frac{1}{\ln t} + \frac{1}{\ln 2}\right] = \frac{1}{\ln 2} < \infty$$

∴ The series converges.

8-52

Determine whether the following series converges or diverges.

$$\sum_{n=1}^{\infty} 3ne^{-n^2}$$

****

$$\int_1^{\infty} 3xe^{-x^2}dx = -\frac{3}{2}e^{-x^2}\Big|_1^{\infty} = \frac{3}{2e}$$

∴ CONVERGES BY THE INTEGRAL TEST.

COMPARISON TESTS

8-53

Determine whether the series $\sum_{n=0}^{\infty} \frac{1 + \sin^2 n}{5^n}$ converges.

****

SINCE $0 \le \sin^2 n \le 1$ FOR ALL n,

$$0 \le \frac{1+\sin^2 n}{5^n} \le \frac{2}{5^n}.$$

SO $$0 \le \sum_{n=0}^{\infty} \frac{1+\sin^2 n}{5^n} \le \sum_{n=0}^{\infty} \frac{2}{5^n}.$$

THE SERIES $\sum_{n=0}^{\infty} \frac{2}{5^n}$ CONVERGES SINCE IT IS A GEOMETRIC SERIES WITH COMMON RATIO $\frac{1}{5}$.

THEREFORE, THE SERIES $\sum_{n=0}^{\infty} \frac{1+\sin^2 n}{5^n}$ CONVERGES BY THE COMPARISON TEST.

8-54

Consider the two series: a) $\sum_{k=2}^{\infty} \frac{\ln k}{k}$ and b) $\sum_{k=2}^{\infty} \frac{1}{k \ln k}$.

Suppose you compare (a) and (b) to the series $\sum_{k=1}^{\infty} \frac{1}{k}$. What (if anything) can you conclude about the convergence or divergence of (a) and (b) using ONLY this comparison test.

****

$\sum_{k=1}^{\infty} \frac{1}{k}$ diverges to $+\infty$

Since $\ln k > 1$ for $k \geq 3$, we have:

a) $\frac{\ln k}{k} > \frac{1}{k}$ and b) $\frac{1}{k \ln k} < \frac{1}{k}$ $(k \geq 3)$

From a) we conclude that $\sum_{k=2}^{\infty} \frac{\ln k}{k}$ also diverges to $+\infty$

b) nothing can be concluded from b) above. the comparison test yields no useful information about the series $\sum_{k=2}^{\infty} \frac{1}{k \ln k}$

8-55

For the series below, tell whether or not it converges, and indicate what test you used. If the test involves a limit, give the limit. If the test involves a comparison, give the comparison.

$$\sum_{n=2}^{\infty} \frac{n^{1/n}}{\ln(n)}$$

****

It diverges:

since $n^{1/n} \geq 1$ for $n \geq 2$ AND $0 < \ln n < n$ for $n \geq 2$, $\frac{n^{1/n}}{\ln n} > \frac{1}{\ln n} > \frac{1}{n}$.

By the comparison test, since $\sum_{n=2}^{\infty} \frac{1}{n}$ diverges, so does the given series.

THE RATIO TEST

8-56

Does the series $\sum_{n=1}^{\infty} \frac{2^{3n}}{5^n}$ converge or diverge? Justify.

****

$$\frac{a_{n+1}}{a_n} = \frac{2^{3(n+1)}}{5^{n+1}} \cdot \frac{5^n}{2^{3n}} = \frac{2^{3n} \cdot 8 \cdot 5^n}{5^n \cdot 5 \cdot 2^{3n}} = \frac{8}{5} > 1$$

∴ by the ratio test this series diverges.

8-57

Test for convergence $\sum_{n=1}^{\infty} \frac{3^n}{n!}$

****

Using the RATIO TEST,

$$\lim_{n\to+\infty} \frac{a_{n+1}}{a_n} = \lim_{n\to+\infty} \left(\frac{\frac{3^{n+1}}{(n+1)!}}{\frac{3^n}{n!}} \right)$$

$$= \lim_{n\to+\infty} \left(\frac{3^{n+1}}{(n+1)!} \cdot \frac{n!}{3^n} \right)$$

$$= \lim_{n\to+\infty} \frac{3^{n+1-n} \cdot n!}{(n+1)(n!)}$$

$$= \lim_{n\to+\infty} \frac{3}{n+1}$$

$$= 0 < 1$$

∴ the series converges.

8-58

Determine if the following series is convergent: $\sum_{n=0}^{\infty} \frac{n!}{3^n}$

****

Use the Ratio Test:

$$r = \lim_{n\to\infty} \left| \frac{a_{n+1}}{a_n} \right| = \lim_{n\to\infty} \frac{\frac{(n+1)!}{3^{n+1}}}{\frac{n!}{3^n}} = \lim_{n\to\infty} \frac{(n+1)!}{3^{n+1}} \cdot \frac{3^n}{n!}$$

$$(n+1)! = (n+1)n! \quad \text{and} \quad 3^{n+1} = 3^n \cdot 3$$

$$r = \lim_{n\to\infty} \frac{(n+1)n!}{3^n \cdot 3} \cdot \frac{3^n}{n!} = \lim_{n\to\infty} \frac{n+1}{3} = \infty .$$

Therefore, $\sum_{n=0}^{\infty} \frac{n!}{3^n}$ is divergent.

MISCELLANEOUS PROBLEMS

8-59

If $\sum_{n=1}^{\infty} a_n = A$ where A is some real number, then $\sum_{n=1}^{\infty} a_n^2$ (a) = A^2 (b) = $|A|$ (c) = $2|A|$ (d) = B where B is unrelated to A (e) may be divergent.

****

Convergence of $\sum_{n=1}^{\infty} a_n$ does not guarantee convergence of $\sum_{n=1}^{\infty} a_n^2$, for example $\sum_{n=1}^{\infty} \frac{(-1)^n}{\sqrt{n}}$ is convergent by the alternating series test, but $\sum_{n=1}^{\infty} \left[\frac{(-1)^n}{\sqrt{n}}\right]^2 = \sum_{n=1}^{\infty} \frac{1}{n}$ is the divergent Harmonic series.

8-60

Find the interval of convergence for the following series: $\sum_{n=0}^{\infty} x^n$.

****

By the ratio test $\sum_{n=0}^{\infty} x^n$ converges when $\lim_{n\to\infty} \left|\frac{x^{n+1}}{x^n}\right| < 1$.

Hence this series converges when $|x| < 1$. So the interval of convergence is the open interval $(-1, 1)$.

8-61

Find the interval of convergence for

$$\sum_{k=0}^{\infty} \left(\frac{e^k}{k+1}\right) x^k$$

****

First apply the ratio test for absolute convergence

$$\lim_{K\to\infty} \left| \frac{e^{K+1} \cdot x^{K+1}}{K+2} \cdot \frac{K+1}{e^K \cdot x^K} \right| = |x|\, e \lim_{K\to\infty} \left(\frac{K+1}{K+2}\right) = |x|\, e$$

Hence the series converges absolutely if

$$|x|\, e < 1 \Rightarrow \frac{-1}{e} < x < \frac{1}{e}$$

Next check for convergence at the endpoints

For $x = -1/e$ we have: $\sum_{K=0}^{\infty} \frac{(-1)^K}{K+1}$, which converges by the alternating series test.

For $x = 1/e$ we have: $\sum_{K=0}^{\infty} \frac{1}{K+1}$, which diverges by limit comparison with the harmonic series $\Sigma 1/K$.

Thus, the interval of convergence for

$$\sum_{K=0}^{\infty} \left(\frac{e^K}{K+1}\right) x^K \quad \text{is} \quad \underline{\underline{\frac{-1}{e} \le x < \frac{1}{e}}}.$$

8-62

Find the interval of convergence of $\Sigma \frac{x^k}{2^k k^2}$.

****

$$\sum \left| \frac{x^k}{2^k k^2} \right| = \sum \frac{|x|^k}{2^k k^2}$$

$$\frac{\dfrac{|x|^{k+1}}{2^{k+1}(k+1)^2}}{\dfrac{|x|^k}{2^k k^2}} = \frac{|x|}{2} \cdot \left(\frac{k}{k+1}\right)^2 \longrightarrow \frac{|x|}{2}$$

IF $\frac{|x|}{2} < 1$ THEN $-2 < x < 2$

$$\sum \frac{(-2)^k}{2^k k^2} = \sum \frac{(-1)^k}{k^2}$$

THIS IS A CONVERGENT ALTERNATING SERIES

$$\sum \frac{2^k}{2^k k^2} = \sum \frac{1}{k^2}$$

THIS IS A CONVERGENT P-SERIES

THUS, THE INTERVAL OF CONVERGENCE IS $[-2,2]$

8-63

Find the interval of convergence for the following infinite series:

$$\sum_{k=1}^{\infty} \frac{(-1)^k (x-3)^k}{5^k (k+1)}$$

****

Use the Ratio Test:

$$\lim_{k\to\infty} \left| \frac{a_{k+1}}{a_k} \right| = \lim_{k\to\infty} \left| \frac{(x-3)^{k+1}}{5^{k+1}(k+2)} \div \frac{(x-3)^k}{5^k(k+1)} \right|$$

$$= \lim_{k\to\infty} \left| \frac{(x-3)(k+1)}{5(k+2)} \right| = \frac{|x-3|}{5}$$

$\therefore$ $|x-3| < 5$ for convergence, and 5 is the radius of convergence.

Check the endpoints:

at $x = 8$, the series becomes $\sum_{k=1}^{\infty} \frac{(-1)^k}{k+1}$,

which is the alternating harmonic series, and therefore converges.

At $x = -2$, the series becomes $\sum_{k=1}^{\infty} \frac{1}{k+1}$,

which clearly diverges.

Hence, the interval of convergence is: $(-2, 8]$

8-64

Analyze each of the following series for convergence vs. divergence:

(a) $\sum_{n=1}^{\infty} \cos(1/n^2)$ (b) $\sum_{n=1}^{\infty} 1/e^n$ (c) $\sum_{n=1}^{\infty} 3^n/n!$ (d) $\sum_{n=1}^{\infty} (\sin n)/n^2$

(e) $\sum_{n=1}^{\infty} n/(5n^2 + \sqrt{n})$ (f) $\sum_{n=1}^{\infty} (-1)^n/n$.

****

(a) $\lim_{n\to\infty} \cos\frac{1}{n^2} = 1$. Therefore $\sum_{n=1}^{\infty} \cos\frac{1}{n^2}$ diverges since $\lim_{n\to\infty} \cos\frac{1}{n^2} \neq 0$.

(b) $\sum_{n=1}^{\infty} \frac{1}{e^n}$ is geometric with $r = \frac{1}{e}$, (so $|r|<1$) and so the series converges.

(c) $\lim_{n\to\infty} \frac{a_{n+1}}{a_n} = \lim_{n\to\infty} \frac{3^{n+1}}{(n+1)!} \cdot \frac{n!}{3^n} = \lim_{n\to\infty} \frac{3}{n+1} = 0 < 1$, thus $\sum_{n=1}^{\infty} \frac{3^n}{n!}$ converges by the ratio test.

(d) $\left|\frac{\sin n}{n^2}\right| \le \frac{1}{n^2}$, and $\sum_{n=1}^{\infty} \frac{1}{n^2}$ is a convergent p-series. Therefore $\sum_{n=1}^{\infty} \left|\frac{\sin n}{n^2}\right|$ is convergent and hence $\sum_{n=1}^{\infty} \frac{\sin n}{n^2}$ is (absolutely) convergent.

(e) Use the limit comparison test and compare to $\sum_{n=1}^{\infty} \frac{1}{n}$, a known divergent series. $\lim_{n\to\infty} \frac{\frac{n}{5n^2+\sqrt{n}}}{\frac{1}{n}} = \lim_{n\to\infty} \frac{n^2}{5n^2+\sqrt{n}}$
$= \lim_{n\to\infty} \frac{1}{5 + \frac{1}{n^{3/2}}} = \frac{1}{5}$. Therefore, $\sum_{n=1}^{\infty} \frac{n}{5n^2+\sqrt{n}}$ is also divergent.

(f) $\sum_{n=1}^{\infty} \frac{(-1)^n}{n}$ is convergent by the alternating series test.

8-65

Find a Taylor series of degree 5 about x = 0 for y = $\mathrm{Tan}^2 x$.

****

Taylor's series about $x=0$ $\quad P_5(x) = f(0) + xf'(0) + \frac{x^2}{2!}f''(0) + \frac{x^3}{3!}f'''(0) + \frac{x^4}{4!}f^{iv}(0) + \frac{x^5}{5!}f^{v}(0)$

We need to determine $f(0), f'(0) \ldots f^{v}(0)$.

$y = \tan^2 x$

$\frac{dy}{dx} = 2\tan x \sec^2 x$

$\frac{d^2y}{dx^2} = 2\tan x(2\sec x . \sec x \tan x) + \sec^2 x(2\sec^2 x) = 4\tan^2 x \sec^2 x + 2\sec^4 x$

now call $y = \tan^2 x$ $\qquad = 4y(1+y^2) + 2(1+y)^2$

$= 2 + 8y + 6y^2$

$\frac{d^3y}{dx^2} = 8\frac{dy}{dx} + 12y\frac{dy}{dx}$

$\frac{d^4y}{dx^4} = 8\frac{d^2y}{dx^2} + 12y\frac{d^2y}{dx^2} + 12\left(\frac{dy}{dx}\right)^2$

$\frac{d^5y}{dx^5} = 8\frac{d^3y}{dx^3} + 12y\frac{d^3y}{dx^3} + 12\left(\frac{dy}{dx}\right)\left(\frac{d^2y}{dx^2}\right) + 24\left(\frac{dy}{dx}\right)\left(\frac{d^2y}{dx^2}\right)$

evaluating $f(0)=0$, $f'(0)=0$, $f''(0)=2$ (since $y=0$) $f'''(0)=0$, $f^{iv}(0)=16$, $f^{v}(0)=0$

hence $P_5(x) = 0 + 0.x + \frac{x^2}{2!}(2) + \frac{x^3}{3!}(0) + \frac{x^4}{4!}(16) + \frac{x^5}{5!}(0)$

So $\tan^2 x \sim P_5(x) = x^2 + \frac{2}{3}x^4$

8-66

Evaluate $\int_0^{.1} \cos x^2 \, dx$ to six decimal places accuracy.

****

$$\cos x = 1 - \frac{x^2}{2!} + \frac{x^4}{4!} - \frac{x^6}{6!} + \cdots$$

$$\therefore \cos x^2 = 1 - \frac{x^4}{2!} + \frac{x^8}{4!} - \frac{x^{12}}{6!} + \cdots$$

$$\therefore \int_0^{.1} \cos x^2 \, dx = \int_0^{.1} \left(1 - \frac{x^4}{2!} + \frac{x^8}{4!} - \frac{x^{12}}{6!} + \cdots\right) dx$$

$$= \left[x - \frac{x^5}{5(2!)} + \frac{x^9}{9(4!)} - \frac{x^{13}}{13(6!)} + \cdots \right]_0^{.1}$$

$$= .1 - \frac{.00001}{10} + \frac{.000000001}{9(4!)} - \cdots$$

$$\doteq 0.099999$$

ONLY TWO TERMS ARE NEEDED BECAUSE THE ERROR FOR A CONVERGENT ALTERNATING SERIES IS LESS THAN THE NEXT TERM.